WJEC GCSE

MATHEMATICS

Higher Homework Book

Wyn Brice, Linda Mason and Tony Timbrell

Second Edition

HODDER
EDUCATION
AN HACHETTE UK COMPANY

The Publishers would like to thank WJEC for permission to reproduce past examination questions throughout this book.

This material has been endorsed by WJEC and offers high quality support for the delivery of WJEC qualifications. While this material has been through a WJEC quality assurance process, all responsibility for the content remains with the publisher.

Hachette UK's policy is to use papers that are natural, renewable and recyclable products and made from wood grown in sustainable forests. The logging and manufacturing processes are expected to conform to the environmental regulations of the country of origin.

Orders: please contact Bookpoint Ltd, 130 Milton Park, Abingdon, Oxon OX14 4SB. Telephone: (44) 01235 827720. Fax: (44) 01235 400454. Lines are open 9.00–5.00, Monday to Saturday, with a 24-hour message answering service. Visit our website at www.hoddereducation.co.uk

© Howard Baxter, Wyn Brice, Mike Handbury, John Jeskins, Jean Matthews, Linda Mason, Mark Patmore, Brian Seager, Tony Timbrell, Eddie Wilde 2010

First published in 2006 by
Hodder Education, an Hachette UK Company,
338 Euston Road
London NW1 3BH

This second edition published in 2010.

Impression number   5 4 3 2
Year   2014 2013

Cover photo © Imagestate Media
Typeset in 10/12pt Times New Roman PS by Pantek Arts Ltd, Maidstone, Kent
Printed by the MPG Printgroup, UK

A catalogue record for this title is available from the British Library

ISBN 978 1444 115307

→ CONTENTS

→ INTRODUCTION

This book contains exercises designed to be used for the Higher tier of GCSE Mathematics. It is particularly aimed at the WJEC Linear and Unitised Specifications and each exercise matches one in the WJEC Higher Student's Book.

In the Homework Book, the corresponding exercises have the same number and end in H. Thus, for example, if you have been working on Further percentages in class and used Exercise 6.1, then the homework exercise is 6.1H. The homework exercises cover the same mathematics.

Some questions are intended to be completed without a calculator, just as in the Student's Book. These are shown with a non-calculator icon in the same way. Doing these questions without a calculator is vital preparation for the GCSE non-calculator paper.

These homework exercises provide extra practice and are also in a smaller book to carry home! If you have understood the topics, you should be able to tackle these exercises confidently as they are no harder than those you have done in class and in some cases may be a little easier. See if you agree.

You will find the answers to this homework book in the Higher Teacher's Website.

1 → INTEGERS, POWERS AND ROOTS

EXERCISE 1.1H

Write each of these numbers as a product of its prime factors.

1	14	2	16
3	28	4	35
5	42	6	49
7	108	8	156
9	225	10	424

EXERCISE 1.2H

For each of these pairs of numbers
- write the numbers as products of their prime factors.
- state the highest common factor.
- state the lowest common multiple.

1	6 and 8	2	8 and 18
3	15 and 25	4	36 and 48
5	25 and 55	6	33 and 55
7	54 and 72	8	30 and 40
9	45 and 63	10	24 and 50

EXERCISE 1.3H

Work out these.

1	2×3	2	-5×8
3	-6×-2	4	-4×6
5	5×-7	6	-3×7
7	-4×-5	8	$28 \div -7$
9	$-25 \div 5$	10	$-20 \div 4$
11	$24 \div 6$	12	$-15 \div -3$
13	$-35 \div 7$	14	$64 \div -8$
15	$27 \div -9$		
16	$3 \times 6 \div -9$		
17	$-42 \div -7 \times -3$		
18	$5 \times 6 \div -10$		
19	$-9 \times 4 \div -6$		
20	$-5 \times 6 \times -4 \div -8$		

EXERCISE 1.4H

 Do not use your calculator for questions **1** and **2**.

1 Write down the value of each of these.
   a) 1^2
   b) 13^2
   c) $\sqrt{64}$
   d) $\sqrt{196}$
   e) 3^3
   f) 5^3
   g) $\sqrt[3]{8}$
   h) $\sqrt[3]{27}$

2 A cube has sides of length 4 cm.
   What is its volume?

 You may use your calculator for questions **3** to **7**.

3 Find the square of each of these numbers.
   a) 20   b) 42   c) 5.1
   d) 60   e) 0.9

4 Find the cube of each of these numbers.
   a) 7   b) 3.5   c) 9.4
   d) 20   e) 100

5 Find the square root of each of these numbers. Where necessary, give your answer correct to 2 decimal places.
   a) 900   b) 75   c) 284
   d) 31 684   e) 40 401

6 Find the cube root of each of these numbers. Where necessary, give your answer correct to 2 decimal places.
   a) 729   b) 144   c) 9.261
   d) 4848   e) 100 000

7 A square has an area of 80 cm².
   What is the length of one side? Give your answer correct to 2 decimal places.

EXERCISE 1.5H

1 Write these in simpler form using indices.
  a) $2 \times 2 \times 2 \times 2 \times 2 \times 2$
  b) $7 \times 7 \times 7 \times 7$
  c) $2 \times 2 \times 3 \times 3 \times 3 \times 3 \times 5 \times 5 \times 5$
    Hint: Write the different numbers separately.

2 Work out these giving your answers in index form.
  a) $2^2 \times 2^4$ **b)** $3^6 \times 3^2$
  c) $4^2 \times 4^3$ **d)** $5^6 \times 5$

3 Work out these giving your answers in index form.
  a) $5^5 \div 5^2$ **b)** $7^8 \div 7^2$
  c) $2^6 \div 2^4$ **d)** $3^7 \div 3^3$

4 Work out these giving your answers in index form.
  a) $5^5 \times 5^3 \div 5^2$ **b)** $10^4 \times 10^6 \div 10^5$
  c) $8^3 \times 8^3 \div 8^4$ **d)** $3^5 \times 3 \div 3^3$

5 Work out these giving your answers in index form.
  a) $\dfrac{2^5 \times 2^4}{2^3}$ **b)** $\dfrac{3^7}{3^5 \times 3^2}$

  c) $\dfrac{5^5 \times 5^4}{5^2 \times 5^3}$ **d)** $\dfrac{7^5 \times 7^2}{7^2 \times 7^4}$

6 **a)** Write each of these numbers as a product of its prime factors.
    (i) 36   **(ii)** 49   **(iii)** 64
    (iv) 100   **(v)** 324
  b) All the numbers in part **a)** are square numbers. Write down what you notice about all the powers (indices) in part **a)**.

7 Write 50 as a product of its prime factors. What is the least number that 50 needs to be multiplied by so that the result is a square number?

8 **a)** Write each of these numbers as a product of its prime factors.
    (i) 12   **(ii)** 27   **(iii)** 60
    (iv) 75   **(v)** 112
  b) For each number in parts **a) (i)** to **(iv)**, find the least number that it needs to be multiplied by so that the result is a square number.

EXERCISE 1.6H

 Do not use your calculator for questions **1** to **3**.

1 Write down the reciprocal of each of these numbers.
  a) 4   **b)** 9   **c)** 65
  d) 10   **e)** 4.5

2 Write down the numbers of which these are the reciprocals.
  a) $\frac{1}{6}$ **b)** $\frac{1}{10}$ **c)** $\frac{1}{25}$
  d) $\frac{1}{71}$ **e)** $\frac{2}{15}$

3 Find the reciprocal of each of these numbers. Give your answers as fractions or mixed numbers.
  a) $\frac{3}{5}$ **b)** $\frac{4}{9}$ **c)** $2\frac{2}{5}$
  d) $5\frac{1}{3}$ **e)** $\frac{3}{100}$

 You may use your calculator for question **4**.

4 Find the reciprocal of each of these numbers. Give your answers as decimals.
  a) 25   **b)** 0.2   **c)** 6.4
  d) 625   **e)** 0.16

2 → FRACTIONS, DECIMALS AND PERCENTAGES

 EXERCISE 2.1H

1 For each pair of fractions
  • find the lowest common denominator.
  • state which is the bigger fraction.

  a) $\frac{7}{8}$ or $\frac{3}{4}$ **b)** $\frac{5}{9}$ or $\frac{7}{11}$ **c)** $\frac{1}{6}$ or $\frac{3}{20}$

2 Work out these.

  a) $\frac{3}{7} + \frac{2}{7}$ **b)** $\frac{7}{15} + \frac{4}{15}$ **c)** $\frac{8}{11} - \frac{3}{11}$
  d) $\frac{11}{17} - \frac{8}{17}$ **e)** $\frac{7}{16} + \frac{3}{16}$ **f)** $\frac{7}{9} + \frac{4}{9}$
  g) $\frac{7}{12} - \frac{5}{12}$ **h)** $\frac{8}{11} + \frac{5}{11}$ **i)** $2\frac{4}{7} + 3\frac{1}{7}$
  j) $4\frac{5}{6} - 1\frac{1}{6}$ **k)** $5\frac{9}{13} - \frac{4}{13}$ **l)** $4\frac{3}{8} - 1\frac{5}{8}$

3 Work out these.

  a) $\frac{2}{9} + \frac{1}{3}$ **b)** $\frac{7}{12} + \frac{1}{4}$ **c)** $\frac{3}{4} - \frac{1}{10}$
  d) $\frac{13}{16} - \frac{3}{8}$ **e)** $\frac{7}{8} + \frac{1}{3}$ **f)** $\frac{4}{5} + \frac{5}{6}$
  g) $\frac{7}{12} - \frac{1}{8}$ **h)** $\frac{9}{20} + \frac{3}{4}$ **i)** $\frac{7}{11} + \frac{3}{5}$
  j) $\frac{7}{12} + \frac{7}{10}$ **k)** $\frac{7}{8} - \frac{1}{6}$ **l)** $\frac{7}{15} - \frac{3}{20}$

4 Work out these.

  a) $4\frac{1}{4} + 3\frac{1}{3}$ **b)** $6\frac{8}{9} - 1\frac{2}{3}$ **c)** $5\frac{3}{8} + \frac{1}{4}$
  d) $5\frac{11}{16} - 2\frac{1}{8}$ **e)** $2\frac{5}{6} + 3\frac{1}{4}$ **f)** $6\frac{8}{9} - 2\frac{1}{6}$
  g) $3\frac{5}{8} + 4\frac{7}{10}$ **h)** $5\frac{7}{11} - 5\frac{1}{3}$ **i)** $4\frac{3}{4} + 3\frac{2}{7}$
  j) $6\frac{1}{4} - 2\frac{2}{3}$ **k)** $7\frac{1}{9} - 2\frac{1}{2}$ **l)** $5\frac{3}{10} - 4\frac{4}{5}$

 EXERCISE 2.2H

1 Change these mixed numbers to improper fractions.

  a) $4\frac{3}{5}$ **b)** $6\frac{1}{4}$ **c)** $3\frac{4}{7}$ **d)** $1\frac{5}{9}$
  e) $4\frac{5}{6}$ **f)** $7\frac{3}{10}$ **g)** $4\frac{7}{8}$

2 Work out these.
Write your answers as proper fractions or mixed numbers in their lowest terms.

  a) $\frac{3}{7} \times 5$ **b)** $\frac{5}{9} \times 6$ **c)** $\frac{3}{5} \div 4$
  d) $6 \times \frac{5}{11}$ **e)** $\frac{2}{9} \div 4$ **f)** $9 \div \frac{3}{8}$

3 Work out these.
Write your answers as proper fractions or mixed numbers in their lowest terms.

  a) $\frac{2}{3} \times \frac{5}{7}$ **b)** $\frac{1}{8} \times \frac{5}{6}$ **c)** $\frac{7}{9} \times \frac{2}{5}$
  d) $\frac{5}{8} \div \frac{3}{4}$ **e)** $\frac{3}{8} \div \frac{1}{3}$ **f)** $\frac{4}{9} \times \frac{5}{11}$
  g) $\frac{6}{7} \times \frac{1}{8}$ **h)** $\frac{7}{15} \div \frac{2}{3}$ **i)** $\frac{7}{12} \times \frac{3}{8}$
  j) $\frac{9}{16} \div \frac{7}{12}$ **k)** $\frac{7}{10} \div \frac{5}{12}$ **l)** $\frac{7}{30} \times \frac{10}{21}$

4 Work out these.
Write your answers as proper fractions or mixed numbers in their lowest terms.

  a) $4\frac{3}{4} \times 1\frac{7}{9}$ **b)** $3\frac{2}{3} \times \frac{1}{5}$ **c)** $4\frac{2}{5} \div 2\frac{4}{5}$
  d) $1\frac{3}{11} \div 3\frac{1}{2}$ **e)** $4\frac{1}{2} \times 3\frac{2}{3}$ **f)** $3\frac{5}{9} \div 2\frac{2}{3}$
  g) $3\frac{2}{7} \times 1\frac{5}{9}$ **h)** $2\frac{5}{8} \div 1\frac{5}{6}$ **i)** $1\frac{7}{15} \times 12\frac{1}{2}$
  j) $5\frac{3}{5} \div 1\frac{3}{4}$ **k)** $6\frac{2}{9} \times 2\frac{1}{8}$ **l)** $7\frac{1}{2} \div 2\frac{3}{5}$

 EXERCISE 2.3H

1 Work out these.

  a) $\frac{3}{4} + \frac{1}{6}$ **b)** $\frac{5}{8} - \frac{2}{7}$
  c) $\frac{5}{9} \times \frac{3}{8}$ **d)** $\frac{7}{16} \div \frac{5}{12}$
  e) $1\frac{4}{5} + 2\frac{3}{4}$ **f)** $6\frac{3}{7} - 2\frac{1}{3}$
  g) $5\frac{3}{5} \times 4$ **h)** $4\frac{5}{9} \div 1\frac{1}{6}$

2 Write these fractions in their lowest terms.

  a) $\frac{40}{125}$ **b)** $\frac{28}{49}$ **c)** $\frac{72}{192}$
  d) $\frac{225}{350}$ **e)** $\frac{17}{153}$

3 Write these improper fractions as mixed numbers.

a) $\frac{120}{72}$ b) $\frac{150}{13}$ c) $\frac{86}{19}$

d) $\frac{192}{54}$ e) $\frac{302}{17}$

4 Calculate
a) the perimeter of this rectangle.
b) the area of this rectangle.

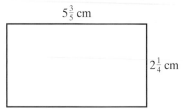

$5\frac{3}{5}$ cm

$2\frac{1}{4}$ cm

 # EXERCISE 2.4H

1 Change each of these fractions to a decimal. If necessary, give your answer to 3 decimal places.

a) $\frac{7}{8}$ b) $\frac{7}{100}$ c) $\frac{5}{9}$ d) $\frac{2}{11}$

2 State whether each of these fractions gives a recurring or a terminating decimal. Give a reason for each answer.

a) $\frac{3}{4}$ b) $\frac{5}{6}$ c) $\frac{5}{11}$

d) $\frac{2}{25}$ e) $\frac{7}{32}$

3 a) Find the recurring decimal equivalent to $\frac{3}{101}$.
b) How many digits are there in the repeating pattern?

 # EXERCISE 2.5H

Work out these. As far as possible, write down only your final answer.

1	3.4 + 6.1	**2**	4.3 + 3.6
3	5.8 − 2.3	**4**	7.9 − 4.4
5	3.7 + 2.6	**6**	5.8 + 3.4
7	7.2 − 0.9	**8**	5.4 − 3.5
9	8.6 + 2.7	**10**	6.9 + 4.6
11	6.7 − 5.8	**12**	6.3 − 2.9

EXERCISE 2.6H

1 Work out these.

a) 5 × 0.4   b) 0.6 × 8
c) 4 × 0.7   d) 0.9 × 6
e) 0.7 × 0.3   f) 0.9 × 0.4
g) 50 × 0.7   h) 0.4 × 80
i) 0.7 × 0.1   j) 0.4 × 0.2
k) $(0.8)^2$ l) $(0.2)^2$

2 Work out these.

a) 6 ÷ 0.3   b) 4.8 ÷ 0.2
c) 2.4 ÷ 0.6   d) 7.2 ÷ 0.4
e) 33 ÷ 1.1   f) 60 ÷ 1.5
g) 12 ÷ 0.4   h) 35 ÷ 0.7
i) 64 ÷ 0.8   j) 32 ÷ 0.2
k) 2.17 ÷ 0.7   l) 47.5 ÷ 0.5

3 Work out these.

a) 3.6 × 1.4   b) 5.8 × 2.6
c) 8.1 × 4.3   d) 6.5 × 3.2
e) 74 × 1.7   f) 64 × 3.8
g) 2.9 × 7.6   h) 11.4 × 3.2
i) 25.2 × 0.8   j) 2.67 × 0.9
k) 8.45 × 1.2   l) 7.26 × 2.4

4 Work out these.

a) 23.6 ÷ 0.4   b) 23.4 ÷ 0.8
c) 18.2 ÷ 0.7   d) 31.2 ÷ 0.6
e) 42.3 ÷ 0.9   f) 75.6 ÷ 1.2
g) 5.28 ÷ 0.3   h) 7.56 ÷ 0.7
i) 63.2 ÷ 0.2   j) 6.27 ÷ 1.1
k) 3.51 ÷ 1.3   l) 8.19 ÷ 1.3

EXERCISE 2.7H

1 Write down the multiplier that will increase an amount by the following percentages.

a) 17%   b) 30%   c) 73%
d) 6%   e) 1%   f) 12.5%
g) 160%

2 Write down the multiplier that will decrease an amount by the following percentages.

a) 13%   b) 40%   c) 35%
d) 8%   e) 4%   f) 27%
g) 15.5%

3 Mrs Green bought an antique for £200.
She later sold it at 250% profit.
What did she sell it for?

4 Jane earns £14 500 per year.
She receives an increase of 2%.
Find her new salary.

5 In a sale, all items are reduced by 20%.
Shamir bought a computer in the sale.
Its original price was £490.
What was its sale price?

6 Graham invested £3500 at 4% compound
interest.
What was the investment worth at the end of
5 years?
Give your answer to the nearest pound.

7 A car decreased in value by 11% per year.
If it cost £16 500 new, what was it worth after
4 years?
Give your answer to a suitable degree of
accuracy.

8 In a certain country, the population rose by
5% every year from 2004 to 2009.
If the population was 26.5 million in 2004,
what was the population in 2009?
Give your answer in millions to the nearest 0.1
of a million.

9 Jane invested £4500 with compound interest
for 3 years.
She could receive either 3% interest every
6 months or 6% interest every year.
Which should Jane choose and
how much more will she receive?

10 Prices went up by 2% in 2007, 3% in 2008
and 2.5% in 2009.
If an item cost £32 at the start of 2007 what
did it cost at the end of 2009?

EXERCISE 3.1H

1 Write each of these ratios in its lowest terms.
   a) 8 : 6   **b)** 20 : 50   **c)** 35 : 55
   d) 8 : 24 : 32   **e)** 15 : 25 : 20

2 Write each of these ratios in its lowest terms.
   a) 200 g : 500 g
   b) 60p : £3
   c) 1 minute : 25 seconds
   d) 2 m : 80 cm
   e) 500 g : 3 kg

3 A bar of brass contains 400 g of copper and 200 g of zinc.
   Write the ratio of copper to zinc in its lowest terms.

4 Teri, Jannae and Abi receive £200, £350 and £450 respectively as their dividends in a joint investment.
   Write the ratio of their dividends in its lowest terms.

5 Three saucepans hold 500 ml, 1 litre and 2.5 litres respectively.
   Write the ratio of their capacities in its lowest terms.

EXERCISE 3.2H

1 Write each of these ratios in the form 1 : n.
   a) 2 : 10   **b)** 5 : 30
   c) 2 : 9   **d)** 4 : 9
   e) 50 g : 30 g   **f)** 15p : £3
   g) 25 cm : 6 m   **h)** 20 : 7
   i) 4 mm : 1 km

2 On a map, a distance of 12 mm represents a distance of 3 km.
   What is the scale of the map in the form 1 : n?

3 A picture is enlarged on a photocopier from 25 mm wide to 15 cm wide.
   What is the ratio of the picture to the enlargement in the form 1 : n?

 ## EXERCISE 3.3H

1 A photo is enlarged in the ratio 1 : 5.
   a) The length of the small photo is 15 cm. What is the length of the large photo?
   b) The width of the large photo is 45 cm. What is the width of the small photo?

2 To make a dressing for her lawn, Rupinder mixes loam and sand in the ratio 1 : 3.
   a) How much sand should she mix with two buckets of loam?
   b) How much loam should she mix with 15 buckets of sand?

3 To make mortar, Fred mixes 1 part cement with 5 parts sand.
   a) How much sand does he mix with 500 g of cement?
   b) How much cement does he mix with 4.5 kg of sand?

4 A rectangular picture is 6 cm wide.
   It is enlarged in the ratio 1 : 4.
   How wide is the enlargement?

5 The Michelin motoring atlas of France has a scale of 1 cm to 2 km.
   a) On the map, the distance between Metz and Nancy is 25 cm.
   How far is the actual distance between the two towns?
   b) The actual distance between Caen and Falaise is 33 km.
   How far is this on the map?

6 Graham is making pastry.
To make enough for five people he uses 300 g of flour.
How much flour should he use to make enough for eight people?

7 To make a solution of a chemical, a scientist mixes 2 parts chemical with 25 parts water.
   a) How much water should he mix with 10 ml of chemical?
   b) How much chemical should he mix with 1 litre of water?

8 The ratio of the sides of two rectangles is 2 : 5.
   a) The length of the small rectangle is 4 cm. How long is the big rectangle?
   b) The width of the big rectangle is 7.5 cm. How wide is the small rectangle?

9 Jason mixes 3 parts black paint with 4 parts white paint to make dark grey paint.
   a) How much white paint does he mix with 150 ml of black paint?
   b) How much black paint does he mix with 1 litre of white paint?

10 In an election, the number of votes was shared between the Labour, Conservative and other parties in the ratio 5 : 4 : 2.
Labour received 7500 votes.
   a) How many votes did the Conservatives receive?
   b) How many votes did the other parties receive?

EXERCISE 3.4H

 Do not use your calculator for questions **1** to **5**.

1 Share £40 between Paula and Tariq in the ratio 3 : 5.

2 Paint is mixed in the ratio 2 parts black paint to 3 parts white paint to make 10 litres of grey paint.
   a) How much black paint is used?
   b) How much white paint is used?

3 A metal alloy is made up of copper, iron and nickel in the ratio 3 : 4 : 2.
How much of each metal is there in 450 g of the alloy?

4 Inderjit worked 6 hours one day.
The time he spent on filing, writing and computing was in the ratio 2 : 3 : 7.
How long did he spend computing?

5 Daisy and Emily invested £5000 and £8000 respectively in a business venture.
They agreed to share the profits in the ratio of their investment.
Emily received £320.
What was the total profit?

 You may use your calculator for questions **6** to **8**.

6 Shahida spends her pocket money on sweets, magazines and clothes in the ratio 2 : 3 : 7.
She receives £15 a week.
How much does she spend on sweets?

7 In a questionnaire the three possible answers are 'Yes', 'No' and 'Don't know'.
The answers from a group of 456 people are in the ratio 10 : 6 : 3.
How many 'Don't knows' are there?

8 Iain and Stephen bought a house between them in Spain.
Iain paid 60% of the cost and Stephen 40%.
   a) Write the ratio of the amounts they paid in its lowest terms.
   b) The house cost 210 000 euros. How much did each pay?

EXERCISE 3.5H

1 An 80 g bag of Munchoes costs 99p and a 200 g bag of Munchoes costs £2.19.
Show which is better value.

2 Baxter's lemonade is sold in 2 litre bottles for £1.29 and in 3 litre bottles for £1.99.
Show which is better value.

3   Butter is sold in 200 g tubs for 95p and in
    450 g packets for £2.10.
    Show which is better value.

4   Fruit yoghurt is sold in packs of 4 tubs for 79p
    and in packs of 12 tubs for £2.19.
    Show which is better value.

5   There are two packs of minced meat on the
    reduced price shelf at the supermarket: a
    1.8 kg pack reduced to £2.50 and a 1.5 kg
    pack reduced to £2.
    Show which is better value.

6   Smoothie shaving gel costs £1.19 for the
    75 ml bottle and £2.89 for the 200 ml bottle.
    Show which is better value.

7   A supermarket sells cans of cola in two
    different sized packs: a pack of 12 cans costs
    £4.30 and a pack of 20 cans costs £7.25.
    Which pack is better value?

8   Sudso washing powder is sold in three sizes:
    750 g for £3.15, 1.5 kg for £5.99 and 2.5 kg
    for £6.99.
    Which packet gives the best value?

4 → MENTAL METHODS

EXERCISE 4.1H

Work these out mentally. As far as possible, write down only the final answer.

1 a) 8 + 25   b) 13 + 49
   c) 0.6 + 5.2   d) 142 + 59
   e) 187 + 25   f) 5.8 + 12.6
   g) 326 + 9.3   h) 456 + 83
   i) 1290 + 41   j) 8600 + 570

2 a) 12 − 5   b) 36 − 9
   c) 10 − 0.4   d) 56 − 45
   e) 82 − 39   f) 141 − 27
   g) 1.2 − 0.7   h) 186 − 19
   i) 307 − 81   j) 1200 − 153

3 a) 7 × 9   b) 12 × 8
   c) 22 × 7   d) 11 × 12
   e) 0.4 × 100   f) 19 × 7
   g) 48 × 5   h) 25 × 8
   i) 63 × 4   j) 42 × 21

4 a) 42 ÷ 7   b) 72 ÷ 12
   c) 1500 ÷ 3   d) 184 ÷ 8
   e) 240 ÷ 20   f) 108 ÷ 18
   g) 96 ÷ 16   h) 16 ÷ 100
   i) 24 ÷ 0.3   j) 3.6 ÷ 0.9

5 a) 7 + (−3)   b) −2 + 6
   c) −5 + 7   d) −4 + (−8)
   e) 4 + (−9)   f) −5 − 1
   g) 6 − (−2)   h) 12 − (−8)
   i) −5 − (−10)   j) −7 − (−3)

6 a) 12 × −2   b) −4 × 5
   c) −6 × −3   d) −10 × −4
   e) 4 × −100   f) 10 ÷ −5
   g) −6 ÷ 2   h) −20 ÷ −4
   i) −35 ÷ −5   j) 32 ÷ −4

7 Write down the square of each of these numbers.
   a) 7   b) 9
   c) 12   d) 14
   e) 1   f) 30
   g) 200   h) 0.5
   i) 0.8   j) 0.2

8 Write down the square root of each of these numbers.
   a) 36   b) 4   c) 64
   d) 121   e) 144

9 Write down the cube of each of these numbers.
   a) 3   b) 4   c) 10
   d) 50   e) 0.2

10 A rectangle has sides 4.6 cm and 5.0 cm.
    Work out
    a) the perimeter of the rectangle.
    b) the area of the rectangle.

11 Ayeesha spends £17.81. How much change does she get from £50?

12 A square has area 121 cm².
    How long is its side?

13 Find 2% of £540.

14 Find two numbers which have a difference of 4 and whose product is 45.

15 Deepa has a 1 kg piece of cheese.
    She eats 432 g of it.
    How much is left?

EXERCISE 4.2H

1   Round each of these numbers to 1 significant figure.
   a) 14.3   b) 38
   c) 6.54   d) 308
   e) 1210   f) 0.78
   g) 0.61   h) 0.053
   i) 2413.5   j) 0.0097

2   Round each of these numbers to 1 significant figure.
   a) 8.4   b) 18.36
   c) 725   d) 8032
   e) 98.3   f) 0.71
   g) 0.0052   h) 0.019
   i) 407.511   j) 23 095

3   Round each of these numbers to 2 significant figures.
   a) 28.7   b) 149.3
   c) 7832   d) 46 820
   e) 21.36   f) 0.194
   g) 0.0489   h) 0.003 61
   i) 0.0508   j) 0.904

4   Round each of these numbers to 3 significant figures.
   a) 7.385   b) 24.81
   c) 28 462   d) 308.61
   e) 16 418   f) 3.917
   g) 60.72   h) 0.9135
   i) 0.004 162   j) 2.236 06

For questions **5** to **12**, round the numbers in your calculations to 1 significant figure.
Show your working.

5   Ashad bought 24 chocolate bars at 32p each. Estimate how much he spent.

6   Ramy earns £382 per week. Estimate his earnings in a year.

7   Mary drove 215 miles in 3 hours 48 minutes. Estimate her average speed.

8   A rectangle has length 9.2 cm and area 44.16 cm². Estimate its width.

9   A new computer is priced at £595 excluding VAT.
    VAT at 17.5% must be paid on it.
    Estimate the amount of VAT to be paid.

10 A square paving slab has area 6000 cm².
    Estimate the length of a side of the slab.

11 A circle has radius 4.3 cm.
    Estimate its area.

12 Estimate the answers to these calculations.
   a) 71×58 b) $\sqrt{46}$
   c) $\dfrac{5987}{5.1}$ d) 19.1^2
   e) 62.7×8316 f) $\dfrac{5.72}{19.3}$
   g) $\dfrac{32}{49.4}$ h) 8152×37
   i) $\dfrac{935 \times 41}{8.5}$ j) $\dfrac{673 \times 0.76}{3.6 \times 2.38}$

EXERCISE 4.3H

Give your answers to these questions as simply as possible.
Leave π in your answers where appropriate.

1 a) $2 \times 6 \times \pi$ b) $\pi \times 7^2$
   c) $\pi \times 12^2$ d) $2 \times 3.8 \times \pi$
   e) $\pi \times 11^2$

2 a) $14\pi + 5\pi$
   b) $\pi \times 3^2 + \pi \times 6^2$
   c) $\pi \times 8^2 - \pi \times 4^2$
   d) $3 \times 42\pi$
   e) $\dfrac{36\pi}{4\pi}$

3   Find the circumference of a circle with radius 15 cm.

4   The areas of two circles are in the ratio $36\pi : 16\pi$. Simplify this ratio.

5   A circular piece of card has radius 12 cm. A square piece is removed, with side 5 cm. Find the area that is left.

EXERCISE 4.4H

1 Work out these.

a) 0.06×600 **b)** 0.03×0.3
c) 0.9×0.04 **d)** $(0.05)^2$
e) $(0.3)^2$ **f)** 500×800
g) 30×5000 **h)** 5.1×300
i) 20.3×2000 **j)** 1.82×5000

2 Work out these.

a) $300 \div 20$ **b)** $60 \div 2000$
c) $3.6 \div 20$ **d)** $1.4 \div 0.2$
e) $2.4 \div 3000$ **f)** $2.4 \div 0.03$
g) $0.08 \div 0.004$ **h)** $5 \div 0.02$
i) $400 \div 0.08$ **j)** $60 \div 0.15$

3 Given that $4.5 \times 16.8 = 75.6$, work out these.

a) 45×1680 **b)** $75.6 \div 168$
c) $7560 \div 45$ **d)** 0.168×0.045
e) $756 \div 0.168$

4 Given that $702 \div 39 = 18$, work out these.

a) $70\,200 \div 39$ **b)** $70.2 \div 3.9$
c) 180×39 **d)** $7.02 \div 18$
e) 1.8×3.9

5 Given that $348 \times 216 = 75\,168$, work out these.

a) $751\,680 \div 216$ **b)** $34\,800 \times 2160$
c) 3.48×21.6 **d)** $751.68 \div 34.8$
e) 0.348×2160

EXERCISE 5.1H

Work these out on your calculator without writing down the answers to the middle stages.

If the answers are not exact, give them correct to 2 decimal places.

1 $\dfrac{7.3 + 8.5}{5.7}$

2 $\dfrac{158 + 1027}{125}$

3 $\dfrac{6.7 + 19.5}{12.2 - 5.7}$

4 $\sqrt{128 - 34.6}$

5 $5.7 + \dfrac{1.89}{0.9}$

6 $(12.6 - 9.8)^2$

7 $\dfrac{8.9}{2.3 \times 5.6}$

8 $\dfrac{15.4}{2.3^2}$

9 $10.9 \times (7.2 - 5.8)$

10 $\dfrac{4.8 + 6.2}{5.2 \times 6.5}$

11 $\dfrac{7.1}{\sqrt{15.3 \times 0.6}}$

12 $\dfrac{3 - \sqrt{2.73 + 5.1}}{4}$

EXERCISE 5.2H

 Do not use your calculator for questions **1** to **4**.

1 These calculations are all wrong. This can be spotted quickly without working them out. For each one, give a reason why it is wrong.
   a) $15.3 \div -5.1 = 5$
   b) $8.7 \times 1.6 = 5.4375$
   c) $4.7 \times 300 = 9400$
   d) $7.5^2 = 46.25$

2 These calculations are all wrong. This can be spotted quickly without working them out. For each one, give a reason why it is wrong.
   a) $5.400 \div 9 = 60$
   b) $-6.2 \times -0.5 = -93.1$
   c) $\sqrt{0.4} = 0.2$
   d) $8.5 \times 7.1 = 60.36$

3 Estimate the answers to each of these calculations. Show your working.
   a) 93×108
   b) 0.61^2
   c) $-19.6 + 5.2$

4 Estimate the answers to these calculations. Show your working.
   a) The cost of three DVDs at £17.99
   b) The cost of 39 cinema tickets at £6.20
   c) The cost of five meals at £7.99 and two drinks at £2.10

 You may use your calculator for questions **5** to **9**.

5 Use inverse operations to check these calculations. Write down the operations you use.
   a) $19\ 669.5 \div 235 = 83.7$
   b) $\sqrt{5069.44} = 71.2$
   c) $9.7 \times 12.4 = 120.28$
   d) $17.2 \times 4.6 + 68.2 = 147.32$

6 Work these out. Round your answers to 2 decimal places.
   a) $\dfrac{24.3 + 18.6}{2.8 \times 0.51}$
   b) $(13.7 + 53.1) \times (9.87 - 5.9)$

7 Work these out. Round your answers to 3 decimal places.
   a) $\dfrac{77.8}{6.4 + 83.9}$
   b) $1.06^4 \times 185$

8 Work these out. Round your answers to 2 significant figures.
   a) $\sqrt{2.5^2 + 9.0}$
   b) 640×0.078

9 a) Use rounding to 1 significant figure to estimate the answer to each of these calculations. Show your working.

   (i) 21.2^3 **(ii)** 189×0.31

   (iii) $\sqrt{11.1^2 - 4.8^2}$ **(iv)** $\dfrac{51.8 + 39.2}{0.022}$

b) Use your calculator to find the correct answer to each of the calculations in part **a)**. Where appropriate, round your answer to a sensible degree of accuracy.

EXERCISE 5.3H

1 Write each of these times as a decimal.
   a) 8 hours 39 minutes
   b) 5 hours 21 minutes
   c) 33 minutes
   d) 3 minutes

2 Write each of these times in hours and minutes.
   a) 4.8 hours   **b)** 5.85 hours
   c) 0.45 hour   **d)** 0.6 hour

3 a) A walker covers a distance of 7.2 miles in 2 hours and 24 minutes.
   Calculate the average speed of the walker in miles per hour.
   b) For one 25-minute section of its journey, an express train travels at an average speed of 126 miles per hour.
   How long is this section of the journey? Give your answer in miles.
   c) In one stage of its flight, a rocket travelled at a constant speed of 357 miles per hour for a distance of 23.8 miles. How long did it take the rocket to travel that distance?

4 In a four-part endurance race, a driver takes the following times to complete each part of the race.

   5 hours 38 minutes
   6 hours 57 minutes
   5 hours 19 minutes
   5 hours 46 minutes

   What was the driver's total time for the race? Give your answer in hours and minutes.

5 Wayne buys some potatoes at £1.25 a kilogram and six nectarines at 37p each. He gives the shop assistant £10 and gets £4.68 change. What weight of potatoes did he buy?

6 Chelsea, Sally and James share the profits from their business in the ratio 4 : 3 : 2. In 2009 the total profit was £94 500. Calculate how much Sally received.

7 Merry followed a recipe for lemon pudding which used 350 g of flour for four people. He made the recipe for 10 people and used a new 1.5 kg bag of flour. How much flour did he have left?

8 Freeville has a population of 36 281 and its area is 27.4 km². Calculate its population density. Give your answer to a sensible degree of accuracy.

9 Mr Brown's mobile phone bill for March showed that he had used 53 minutes of calls at 13p per minute. His monthly rental charge for the phone was £15.30. There was VAT at 20% on the whole bill. Calculate the total bill including VAT.

10 In January 2009, the Retail Price Index (RPI) was 210.1. In January 2010 it was 217.9. Calculate the percentage increase in the RPI over that year.

11 A metal cylinder has radius 3.8 cm and height 5.7 cm. Its density is 12 g/cm³. Calculate its mass.

12 A water urn is in the shape of a cuboid, with the base a square of side 30 cm. How many litres of water does the urn contain when it is filled to a depth of 42 cm?

13 David drove 28 miles along the motorway at 70 mph, and then 10 miles at 50 mph.
   a) Calculate how long he took.
   b) Find his average speed for the whole journey.

EXERCISE 6.1H

1   Calculate how much these items are worth if they increase by the given fraction each year for the given number of years.
    Give your answers to the nearest penny.

	Original value	Fractional increase	Number of years
a)	£5000	$\frac{1}{30}$	4
b)	£300	$\frac{2}{7}$	3
c)	£4500	$\frac{5}{9}$	6

2   Calculate how much these items are worth if they decrease by the given fraction each year for the given number of years.
    Give your answers to the nearest penny.

	Original value	Fractional decrease	Number of years
a)	£120	$\frac{1}{6}$	5
b)	£5200	$\frac{3}{5}$	6
c)	£140	$\frac{1}{2}$	4

3   Cathy invested £1870 with a bond that offered to increase the amount by $\frac{1}{12}$ each year.
    How much was the bond worth after 4 years?

4   Patnik shoe shop reduced the price of a pair of shoes by $\frac{1}{4}$ each day until they were sold.
    The original price of the shoes was £47.
    What was the price after 4 days?
    Give your answer to the nearest penny.

5   It is estimated that the royalties from a book will decrease by $\frac{2}{5}$ each year.
    In 2009 Kath received £11 000 in royalties. If the estimate is correct, how much will she receive 5 years later?

EXERCISE 6.2H

1   Copy and complete this table.

	Original value	Percentage increase	Increased value
a)	£750	8%	
b)		15%	£414
c)	£42.50	4.5%	
d)		5%	£254

2   Copy and complete this table.

	Original value	Percentage decrease	Decreased value
a)	£2000	12%	
b)		5%	£240
c)	£260	3.5%	
d)		12.5%	£325

3   After an increase of 12%, a quantity is 84 tonnes.
    What was it before the increase?

4   A newspaper increased its circulation by 3% and the new number sold was 58 195.
    What was the circulation before the increase?

5   Santos bought a car for £14 750.
    He sold it three years later at a loss of 45%.
    What price did he sell it for?

6   A charity's income has been reduced by 2.5%.
    Its income is now £8580.
    What was it before the reduction?

7   It was announced that the number of people
    unemployed had decreased by 3%.
    The number unemployed before the decrease
    was 2.56 million.
    How many are now unemployed?

8   The price of a car is £13 200, including VAT
    at 20%.
    What is the price without VAT?

9   At Percival's sale the price of everything was
    reduced by 7.5%, rounded to the nearest
    penny.
    a) A pair of boots cost £94.99 before the sale.
        What was the price in the sale?
    b) Delia was charged £13.87 for a blouse in
        the sale.
        What was its original price?

10 John's pension has increased by 4.75% and is
    now £924.56 a month.
    What was it before the increase?

11 At St Bede's church they give $\frac{1}{8}$ of their
    weekly collection to overseas aid.
    One week they had £151.20 left after they had
    given away the money.
    How much was the total collection?

12 The makers of Steamer's jam say that the new
    size jar contains a fifth more jam than the old
    size jar.
    The new size jar contains 570 g.
    How much did the old size jar contain?

EXERCISE 7.1H

1 Write these in index form.
   a) $\sqrt[5]{x}$
   b) The reciprocal of $x^{\frac{1}{2}}$
   c) $\sqrt[4]{x^3}$

2 Work out these. Give your answers as whole numbers or fractions.
   a) 25^{-1} b) $25^{\frac{1}{2}}$ c) 25^0
   d) $25^{-\frac{1}{2}}$ e) $25^{\frac{3}{2}}$ f) $27^{\frac{2}{3}}$
   g) $10\,000^{\frac{1}{4}}$ h) $\left(\frac{1}{100}\right)^{-\frac{1}{2}}$ i) $32^{\frac{6}{5}}$
   j) $\left(\frac{1}{2}\right)^0$ k) $16^{\frac{1}{2}}$ l) 16^0
   m) $16^{\frac{3}{2}}$ n) $16^{-\frac{1}{4}}$ o) $16^{\frac{7}{4}}$
   p) $144^{\frac{1}{2}}$ q) $\left(\frac{2}{5}\right)^{-2}$

3 Work out these. Give your answers as whole numbers or fractions.
   a) $1000^{\frac{2}{3}} \times 8^{\frac{2}{3}}$
   b) $100^{-\frac{1}{2}} \times 49^{\frac{3}{2}}$
   c) $4^{-2} \times 10^4 \times 25^{-\frac{1}{2}}$
   d) $4^3 + 16^{\frac{1}{2}} - \left(\frac{1}{5}\right)^{-2}$
   e) $10\,000^{\frac{1}{4}} + 125^{\frac{1}{3}} - 121^{\frac{1}{2}}$
   f) $\left(\frac{4}{5}\right)^2 \times 128^{-\frac{3}{7}}$

 EXERCISE 7.2H

1 Work out these. Give your answers exactly or to 5 significant figures.
   a) 5.3^4 b) 0.72^5
   c) 1.03^7 d) 1.37^{-3}
   e) $28\,561^{\frac{1}{4}}$ f) $9.23^{\frac{1}{5}}$
   g) $\sqrt[4]{93}$ h) $2187^{\frac{3}{7}}$

2 Work out these. Give your answers exactly or to 5 significant figures.
   a) 300×1.05^{10} b) $2.4^6 \times 1.2^5$
   c) $5.7^3 \div 2.6^{-3}$ d) $(4.7 \times 5.1^4)^{\frac{1}{5}}$
   e) $3.7^4 + 1.3^7$ f) $3.7^{\frac{1}{4}} - 12.8^{\frac{1}{6}}$
   g) $2.7^5 + 0.8^{-4}$ h) $512^{\frac{4}{3}} \div 3125^{\frac{2}{5}}$

 EXERCISE 7.3H

1 Write these as powers of 2 as simply as possible.
   a) 64   b) $8^{\frac{2}{3}}$ c) 0.25
   d) $2 \times \sqrt[3]{64}$ e) $4^{\frac{1}{2}}$ f) $2^{3n} \times 4^{\frac{n}{2}}$

2 Where possible, write these as powers of a prime number as simply as possible.
   a) 343   b) $9^{\frac{2}{3}}$
   c) $64^{-\frac{2}{3}}$ d) $16^{\frac{1}{2}} \times 64^{-\frac{2}{3}}$
   e) $2^6 + 2^3$ f) $27 \div 81^{\frac{3}{2}}$
   g) $9^{3n} \times 3^{-n}$

3 Write each of these in the form $2^a \times 3^b$ or $2^a \times 3^b \times 5^c$.
   a) 60   b) 192
   c) 600   d) 648

4 Write each of these numbers as a product of powers of prime numbers.
   a) 15^3 b) $25^2 \times 10^{\frac{1}{2}}$
   c) 40^n d) $40^n \times 10$

EXERCISE 7.4H

1   Write these numbers in standard form.
   a)   60 000   b)   8400
   c)   863 000   d)   72 500 000
   e)   9 020 000   f)   78
   g)   5.2 million

2   Write these numbers in standard form.
   a)   0.08   b)   0.0096
   c)   0.000 308   d)   0.000 063
   e)   0.000 004 8   f)   0.000 000 023

3   These numbers are in standard form. Write
    them as ordinary numbers.
   a) 3×10^3 b) 4.6×10^4
   c) 2×10^{-5} d) 7.2×10^5
   e) 1.9×10^{-4} f) 5.78×10^6
   g) 2.87×10^8 h) 5.13×10^{-6}
   i) 2.07×10^{-3} j) 7.28×10^7

EXERCISE 7.5H

1   Work out these. Give your answers in standard
    form.
   a) $(6 \times 10^4) \times (3 \times 10^5)$
   b) $(8 \times 10^6) \div (4 \times 10^3)$
   c) $(9 \times 10^5) \times (4 \times 10^{-2})$
   d) $(3 \times 10^{-3}) \times (2 \times 10^{-4})$
   e) $(2 \times 10^8) \div (5 \times 10^3)$
   f) $(7.6 \times 10^5) + (3.8 \times 10^4)$
   g) $(5.6 \times 10^{-3}) - (4 \times 10^{-4})$

2   Work out these. Give your answers in standard
    form.
   a) $(4.4 \times 10^6) \times (2.7 \times 10^5)$
   b) $(6.5 \times 10^6) \times (2.3 \times 10^2)$
   c) $(7.1 \times 10^4) \times (8.3 \times 10^2)$
   d) $(2.82 \times 10^4) \div (1.2 \times 10^{-2})$
   e) $(7.2 \times 10^3) \times (1.3 \times 10^5)$
   f) $(4.3 \times 10^3) + (6.72 \times 10^4)$
   g) $(6.21 \times 10^5) - (3.75 \times 10^4)$

EXERCISE 8.1H

Use your calculator to work out these.
Give all your answers to 3 significant figures.

1　**a)** $\dfrac{1}{1.7} + \dfrac{1}{1.653}$

　　b) $\dfrac{1}{0.9} \times \dfrac{1}{8.24}$

　　c) $\dfrac{8.06}{5.91} - \dfrac{1.594}{1.62}$

2　**a)** 0.741^3　　**b)** 2.28^{-5}　**c)** $(9.2 + 15.3)^2$

3　**a)** $\sqrt[4]{12.2}$

　　b) $\sqrt[3]{0.8145 - 0.757}$

　　c) $\sqrt[5]{8.6^2 + 9.71^3}$

4　**a)** $\sin 14.6°$

　　b) $\tan 71.3°$

　　c) $\sin 247° - \cos(-31)°$

5　**a)** $\cos^{-1} 0.141$

　　b) $\sin^{-1} 0.464$

　　c) $\tan^{-1} \dfrac{1}{\sqrt{2}}$

6　**a)** $(1.2 \times 10^4) \times (5.3 \times 10^6)$

　　b) $\dfrac{4.06 \times 10^{-2}}{7 \times 10^{-4}}$

　　c) $(5.9 \times 10^{-3}) \times (2.4 \times 10^{20})$

7　**a)** $\dfrac{9.71 \times 0.008\ 476\ 5}{5.9^2}$

　　b) $\dfrac{81.7 + 1.52}{62.8}$

　　c) $\dfrac{9.61}{17.37 \times 224}$

8　**a)** $\dfrac{101}{27.4 + 296}$

　　b) $\dfrac{3.14 \times 0.782}{22.4 - 15.5}$

　　c) $\dfrac{18.21 - 5.63}{23.48 + 19.76}$

9　**a)** $3 \cos 12° - 5 \sin 12°$

　　b) $\dfrac{3.4 \times \sin 47.1°}{\sin 19.2°}$

　　c) $2.7^2 + 3.6^2 - 2 \times 2.7 \times 3.6 \cos 25°$

10　**a)** $(3.6 \times 10^{-8})^2$

　　b) $\sqrt{4.84 \times 10^6}$

　　c) $\dfrac{(4.06 \times 10^6) + (1.15 \times 10^7)}{5.83 \times 10^{-6}}$

EXERCISE 8.2H

1　£30 000 is invested at 18% compound interest.
　　a) Write down a formula for the amount, a, the investment is worth after t years.
　　b) Calculate the value of the investment after
　　　　(i) 5 years.　　　**(ii)** 12 years.

2　The value of a currency, the scud, depreciates by 6% each month.
　　I have 250 000 scuds.
　　a) Write down a formula for the value of the currency, c, after t months.
　　b) Calculate the value of 250 000 scuds after
　　　　(i) 4 months.　　　**(ii)** 9 months.
　　c) Use trial and improvement to work out how many months it will be before the 250 000 scuds are only worth as much as 100 000 scuds.

3 The population of a rare species of animal is decaying exponentially following the formula

$$N = 60\,000 \times 2^{-t}$$

where N is the number of these animals present and t is the time in years.
a) How many of these animals were there when the survey started?
b) How many of these animals were there after
   (i) 3 years? (ii) 10 years?
c) After how many years will this species of animal no longer exist; that is, after how many years will the number fall below 2?

4 Liz's employer tells her that her weekly wage will increase by a fixed percentage each year. They tell her that the formula they will use is

$$A = 500 \times 1.04^n.$$

a) How much is Liz's current weekly wage?
b) What is the rate of increase?
c) What does the letter n stand for in the formula?
d) How much will her wage be after
   (i) 4 years? (ii) 10 years?

5 The population of a country is increasing at a rate of 3% per year.
In 2010 the population was 38 million.
a) Write down a formula for the population size, P, after t years.
b) What will be the population in
   (i) 2015? (ii) 2105?
c) How long will it take for the population to double from its 2010 size?

6 A sample of a radioactive element has a mass of 1 kg.
Its mass reduces by 10% each year.
a) Write down a formula for the mass, m, of the element after t years.
b) Calculate the mass after
   (i) 4 years. (ii) 8 years.
c) Use trial and improvement to find how long it takes for the mass to halve.

7 a) Draw the graph of $y = 3^x$ for values of x from 0 to 3.

b) Use your graph to find
   (i) the value of y when $x = 1.5$.
   (ii) the solution of the equation $3^x = 15$.

8 a) Draw the graph of $y = 2^{-x}$ for values of x from -4 to 0.
b) Use your graph to find
   (i) the value of y when $x = -2.5$.
   (ii) the solution of the equation $2^{-x} = 10$.

EXERCISE 8.3H

1 Find the upper and lower bounds for each of these measurements.
a) The height of a house is 8.5 m, to the nearest 0.1 m.
b) A child weighs 57 kg, to the nearest kilogram.
c) A door is 0.75 m wide, to the nearest centimetre.
d) The winning time for the 100 m race was 12.95 seconds, to the nearest hundredth of a second.
e) The volume of milk in a bottle is 500 ml, to the nearest millilitre.

2 A garden is measured as 43 m, to the nearest metre.
Write the possible length of the garden as an inequality.

3 A kitchen unit is 60 cm wide, to the nearest centimetre.
a) Is it possible that the unit will fit into a gap 60 cm wide, to the nearest centimetre? Show how you decide.
b) Is it certain that the unit will fit into the gap? Show how you decide.

4 Find the upper and lower bounds for each of these measurements.
a) The total weight of 10 books, each book weighing 1.5 kg to the nearest 0.1 kg.
b) The total length of 20 paper clips, each measuring 3 cm to the nearest centimetre.
c) The total time to make 100 sandwiches when each sandwich takes 40 seconds to make, to the nearest second.

EXERCISE 8.4H

1 The lengths of the sides of a rectangle are 8 cm and 10 cm.
Both measurements are correct to the nearest centimetre.
Work out the upper and lower bounds of the perimeter of the rectangle.

2 Work out the largest and smallest possible areas of a triangle with a base of 7 cm and a height of 5 cm, where both lengths are correct to the nearest centimetre.

3 Sugar weighing 0.1 kg is taken from a bag weighing 2 kg.
Both weights are correct to the nearest 0.1 kg.
What are the maximum and minimum possible weights of the sugar remaining?

4 Two stages of a relay race are run in times of 14.07 seconds and 15.12 seconds, to the nearest 0.01 second.
Calculate the upper bound of
a) the total time of the two stages.
b) the difference between the times for the two stages.

5 Given that $p = 5.1$ and $q = 8.6$, correct to 1 decimal place, work out the largest and smallest possible values of
a) $p \times q$.   b) $q \div p$.

6 A train travels 150 miles in 1.8 hours.
The distance is correct to the nearest mile and the time is correct to the nearest 0.1 hour.
Work out the upper and lower bounds of the speed of the train.

7 The density of an object is given as 5.7 g/cm³, to the nearest 0.1 g/cm³.
Its volume is 72.5 cm³, to 3 significant figures.
Find the upper and lower bounds of the mass of the object.

8 Scott is trying to work out a value for π.
He measures the circumference of a circle as 32 cm and the diameter as 10 cm, both correct to the nearest centimetre.
Calculate the upper and lower bounds of Scott's value for π.

9 Use the formula $a = \dfrac{v^2}{2s}$ to work out the upper and lower bounds of a when $v = 2.1$ and $s = 5.7$ and both values are correct to 1 decimal place.

10 Work out the upper and lower bounds of the calculation
$$\frac{9.4 - 5.2}{3.8}$$
if each value in the calculation is correct to 2 significant figures.

11 Concrete blocks have a mass of 15 kg measured to the nearest kg.
a) Write down the least and greatest possible values of the mass of a concrete block.
b) (i) Find the least and greatest possible values of the mass of 100 concrete blocks.
(ii) Denver wishes to be sure that he puts no more than 1500 kg of blocks on his lorry. Find the greatest number of blocks Denver should put on his lorry in order to be sure that no more than 1500 kg is loaded.

12 A jug has a volume of 500 cm³, measured to the nearest 10 cm³.
a) Write down the least and greatest possible values of the volume of the jug.
b) Water is poured from the jug into a tank of volume 15.5 litres measured to the nearest 0.1 litre.
Explain, showing all your calculations, why it is always possible to pour water from 30 full jugs into the tank without it overflowing.

9 → DECIMALS AND SURDS

EXERCISE 9.1H

1 Which of these fractions are equivalent to recurring decimals?

   a) $\frac{1}{8}$ **b)** $\frac{7}{30}$ **c)** $\frac{4}{7}$

   d) $\frac{5}{16}$ **e)** $\frac{11}{120}$

2 Find the decimal equivalent of each of the fractions in question **1**.

3 When these fractions are written as decimals, which of them terminate?

   a) $\frac{6}{40}$ **b)** $\frac{7}{8}$ **c)** $\frac{1}{6}$

   d) $\frac{2}{75}$ **e)** $\frac{11}{80}$

4 Find the decimal equivalent of each of the fractions in question **3**.

5 Find the fractional equivalent of each of these terminating decimals.
Write each fraction in its simplest form.

   a) 0.16   **b)** 0.305

   c) 0.625   **d)** 0.408

6 Find the fractional equivalent of each of these recurring decimals.
Write each fraction in its simplest form.

   a) $0.\dot{3}$ **b)** $0.\dot{8}$ **c)** $0.1\dot{5}$ **d)** $0.7\dot{2}$

7 Find the fractional equivalent of each of these recurring decimals.
Write each fraction in its simplest form.

   a) $0.\dot{6}\dot{3}$ **b)** $0.\dot{4}\dot{7}$ **c)** $0.1\dot{5}$ **d)** $0.3\dot{8}$

8 Find the fractional equivalent of each of these recurring decimals.
Write each fraction in its simplest form.

   a) $0.3\dot{0}\dot{6}$ **b)** $0.\dot{4}1\dot{4}$

   c) $0.0\dot{8}$ **d)** $0.02\dot{7}$

EXERCISE 9.2H

1 Simplify these.

   a) $\sqrt{7} + 3\sqrt{7}$

   b) $9\sqrt{5} - 2\sqrt{5}$

   c) $4\sqrt{3} - \sqrt{3}$

   d) $\sqrt{3} \times 5\sqrt{3}$

   e) $6\sqrt{2} \times \sqrt{5}$

   f) $3\sqrt{3} \times \sqrt{12}$

2 Write each of these expressions in the form $a\sqrt{b}$, where b is an integer which is as small as possible.

   a) $\sqrt{18}$

   b) $3\sqrt{50}$

   c) $5\sqrt{12}$

   d) $6\sqrt{20}$

   e) $\sqrt{108}$

   f) $\sqrt{1250}$

3 Simplify these.

   a) $\sqrt{18} + 4\sqrt{2}$

   b) $7\sqrt{3} - \sqrt{75}$

   c) $\sqrt{12} + 3\sqrt{3}$

   d) $\sqrt{8} \times 3\sqrt{6}$

   e) $\sqrt{21} \times \sqrt{12}$

   f) $2\sqrt{250} \times \sqrt{10}$

4 Expand and simplify these.

   a) $\sqrt{2}(5 + \sqrt{2})$

   b) $\sqrt{3}(\sqrt{12} + \sqrt{2})$

   c) $4\sqrt{5}(2 + \sqrt{45})$

   d) $3\sqrt{2}(\sqrt{18} + \sqrt{5})$

5 Expand and simplify these. State whether your answer is rational or irrational.

a) $(1 + 2\sqrt{7})(3 + 4\sqrt{7})$

b) $(3 - \sqrt{3})(8 + \sqrt{3})$

c) $(4 - \sqrt{5})(3 - 2\sqrt{5})$

d) $(\sqrt{5} + 1)(\sqrt{5} - 4)$

e) $(\sqrt{10} + 5)(\sqrt{10} - 5)$

f) $(6 - \sqrt{13})(4 - 2\sqrt{13})$

6 Find the value of each of these expressions when $m = 4 + \sqrt{5}$ and $n = 6 - 2\sqrt{5}$.

a) $3n$ **b)** $m + 2n$

c) $3m - 2n$ **d)** mn

7 Find the value of each of these expressions when $p = 7 + 2\sqrt{3}$ and $q = 7 - 2\sqrt{3}$.

a) $4p$ **b)** $\sqrt{3}q$

c) $p - q$ **d)** pq

e) p^2 **f)** q^2

8 Given that $r = 7 + \sqrt{5}$ and $s = 6 - 4\sqrt{5}$

a) find the value of k if $3r + ks$ is an integer.

b) find the value of k if $3r + ks$ is of the form $a\sqrt{5}$.

9 Rationalise the denominator and simplify each of these.

a) $\dfrac{15}{\sqrt{3}}$ **b)** $\dfrac{4}{\sqrt{5}}$

c) $\dfrac{6}{5\sqrt{2}}$ **d)** $\dfrac{6}{7\sqrt{12}}$

e) $\dfrac{6\sqrt{3}}{5\sqrt{2}}$ **f)** $\dfrac{12\sqrt{10}}{11\sqrt{6}}$

10 Rationalise the denominator and simplify each of these.

a) $\dfrac{10 + 2\sqrt{5}}{\sqrt{5}}$

b) $\dfrac{12 + \sqrt{3}}{4\sqrt{3}}$

c) $\dfrac{6 + 5\sqrt{2}}{\sqrt{2}}$

d) $\dfrac{5 + 10\sqrt{3}}{\sqrt{15}}$

10 → ALGEBRAIC MANIPULATION 1

EXERCISE 10.1H

Expand these.

1 $7(3a + 6b)$
2 $5(2c + 3d)$
3 $4(3e - 5f)$
4 $3(7g - 2h)$
5 $3(4i + 2j - 3k)$
6 $3(5m - 2n + 3p)$
7 $6(4r - 3s - 2t)$
8 $8(4r + 2s + t)$
9 $4(3u + 5v)$
10 $6(4w + 3x)$
11 $2(5y + z)$
12 $4(3y + 2z)$
13 $5(3v + 2)$
14 $3(7 + 4w)$
15 $5(1 - 3a)$
16 $3(8g - 5)$

EXERCISE 10.2H

Expand the brackets and simplify these.

1 a) $3(4a + 5) + 2(3a + 4)$
   b) $5(4b + 3) + 3(2b + 1)$
   c) $2(3 + 6c) + 4(5 + 7c)$
2 a) $2(4x + 5) + 3(5x - 2)$
   b) $4(3y + 2) + 5(3y - 2)$
   c) $3(4 + 7z) + 2(3 - 5z)$
3 a) $4(4s + 3t) + 5(2s + 3t)$
   b) $3(4v + 5w) + 2(3v + 2w)$
   c) $6(2x + 5y) + 3(4x + 2y)$
   d) $2(5v + 4w) + 3(2v + w)$
4 a) $5(2n + 5p) + 4(2n - 5p)$
   b) $3(4q + 6r) + 5(2q - 3r)$
   c) $7(3d + 2e) + 5(3d - 2e)$
   d) $5(3f + 8g) + 4(3f - 9g)$
   e) $4(5h - 6j) - 6(2h - 5j)$
   f) $4(5k - 6m) - 3(2k - 5m)$

EXERCISE 10.3H

Factorise these as fully as possible.

1 a) $8x + 20$
   b) $3x + 6$
   c) $9x - 12$
   d) $5x - 30$
2 a) $16 + 8x$
   b) $9 + 15x$
   c) $12 - 16x$
   d) $8 - 12x$

3 a) $4x^2 + 16x$
   b) $6x^2 + 30x$
   c) $8x^2 - 20x$
   d) $9x^2 - 15x$

EXERCISE 10.4H

Expand the brackets and simplify these.

1 a) $(a + 5)(a + 3)$
   b) $(b + 2)(b + 4)$
   c) $(4 + c)(3 + c)$
2 a) $(2d + 3)(4d - 3)$
   b) $(5e + 4)(3e - 2)$
   c) $(3 + 8f)(2 - 5f)$
3 a) $(3g - 2)(5g - 6)$
   b) $(4h - 5)(3h - 7)$
   c) $(5j - 6)(3j - 8)$
4 a) $(3k + 7)(4k - 5)$
   b) $(2 + 7m)(3 - 8m)$
   c) $(4 + 3n)(2 - 5n)$
5 a) $(2 + 5p)(3 - 7p)$
   b) $(5r - 6)(2r - 5)$
   c) $(3s - 2)(4s - 9)$

EXERCISE 10.5H

Simplify each of the following, writing your answer using index notation.

1 a) $7 \times 7 \times 7 \times 7 \times 7$
   b) $3 \times 3 \times 3 \times 3 \times 3$
   c) $2 \times 2 \times 2 \times 2 \times 2 \times 2$
2 a) $d \times d \times d \times d \times d \times d \times d$
   b) $m \times m \times m \times m \times m \times m$
   c) $t \times t \times t \times t \times t \times t \times t$
3 a) $a \times a \times a \times a \times b \times b$
   b) $c \times c \times c \times c \times d \times d \times d \times d \times d$
   c) $r \times r \times r \times s \times s \times t \times t \times t \times t$
4 a) $2x \times 3y \times 6z$
   b) $2a \times 3b \times 4c$
   c) $r \times 2s \times 3t \times 4s \times 5r$

11 → ALGEBRAIC MANIPULATION 2

EXERCISE 11.1H

Simplify each of these algebraic fractions.

1. $\dfrac{7x - 14}{4x + 8}$

2. $\dfrac{3x - 6}{9 - 12x}$

3. $\dfrac{12 + 8x}{6x - 4}$

4. $\dfrac{18 - 6x}{12 + 9x}$

5. $\dfrac{4x^2 + 2x}{2x^2 - 6x}$

6. $\dfrac{3x^2 - 4x}{5x^2 - 6x}$

7. $\dfrac{8 - 6x}{4x^2 + 6x}$

8. $\dfrac{3x^2 + 6x}{9x - 12x^2}$

EXERCISE 11.2H

Factorise each of these expressions.

1. $x^2 + 7x + 6$
2. $x^2 + 8x + 15$
3. $x^2 + 9x + 18$
4. $x^2 + 14x + 13$
5. $x^2 + 10x + 25$
6. $x^2 + 11x + 24$
7. $x^2 + 13x + 40$
8. $x^2 + 19x + 48$

EXERCISE 11.3H

Factorise each of these expressions.

1. $x^2 - 1$
2. $x^2 - 36$
3. $x^2 - 121$
4. $x^2 - 196$
5. $2x^2 - 32$
6. $4x^2 - 100$
7. $6x^2 - 54$
8. $9x^2 - 576$

EXERCISE 11.4H

Simplify each of these expressions.

1. $\dfrac{8a^3b^5}{9c^4} \times \dfrac{15c^2}{4a^5b^3}$

2. $\dfrac{15x^3y^2}{16z} \times \dfrac{12z^3}{25x^4y^5}$

3. $\dfrac{4p^5}{3r^6s^7} \times \dfrac{15r^6s^4}{8p^2}$

4. $\dfrac{7h^4j^2}{6k} \div \dfrac{14h^7j^5}{15k^3}$

5. $\dfrac{20t^6}{9v^5w^2} \div \dfrac{15t^2}{6v^3w^4}$

6. $\dfrac{18a^8b^2}{25c^5} \div \dfrac{27a^5b^4}{20c^9}$

7. $\dfrac{15a^4b^5}{8c^4d^3} \times \dfrac{24b^5c^4}{35a^2d^3} \times \dfrac{7c^3d^2}{6ab^6}$

8. $\dfrac{9w^6x^5}{10yz^2} \div \dfrac{3w^3y^4}{4x^4z^3} \times \dfrac{5w^2z}{6x^5y^6}$

EXERCISE 11.5H

Factorise each of these expressions.

1 $3x^2 + 17x + 10$

2 $2x^2 + 7x + 3$

3 $2x^2 + x - 21$

4 $5x^2 - 19x - 4$

5 $2x^2 + x - 15$

6 $3x^2 + 11x - 20$

7 $2x^2 - 11x + 12$

8 $5x^2 - 19x + 12$

9 $6x^2 + 17x + 5$

10 $10x^2 - 21x - 10$

11 $12x^2 - 17x + 6$

12 $21x^2 + 19x - 12$

EXERCISE 11.6H

Simplify each of these algebraic fractions.

1 $\dfrac{x + 4}{x^2 + 2x - 8}$

2 $\dfrac{x^2 - 2x - 15}{x - 5}$

3 $\dfrac{x^2 - 7x + 12}{x^2 - x - 12}$

4 $\dfrac{x^2 - 2x - 8}{x^2 - 6x + 8}$

5 $\dfrac{3x^2 + 7x + 2}{x^2 - 7x + 6}$

6 $\dfrac{6x^2 + 8x - 8}{3x^2 + 7x - 6}$

7 $\dfrac{x^2 + x - 2}{x^2 - 1}$

8 $\dfrac{x^2 + x - 12}{x^2 - 16}$

9 $\dfrac{4x + 8}{x^2 + 6x + 8}$

10 $\dfrac{10x + 15}{4x^2 + 6x}$

11 $\dfrac{x^2 + 2x - 8}{x^2 - 5x + 6}$

12 $\dfrac{x^2 - 2x - 8}{x^2 - 16}$

13 $\dfrac{12x^2 + 15x}{4x^2 - 7x - 15}$

14 $\dfrac{3x^2 + 14x + 8}{6x^2 - 5x - 6}$

15 $\dfrac{3(x - 4)^2}{4x^2 - 64}$

16 $\dfrac{3x(x + 4)^2}{x^2 + x - 12}$

12 → EQUATIONS AND INEQUALITIES 1

EXERCISE 12.1H

Solve these equations.

1 $2x - 3 = 7$
2 $2x + 2 = 8$
3 $2x - 9 = 3$
4 $3x - 2 = 7$
5 $6x + 2 = 26$
6 $3x + 2 = 17$
7 $4x - 5 = 3$
8 $4x + 2 = 8$
9 $2x - 7 = 10$
10 $5x + 12 = 7$
11 $x^2 + 3 = 19$
12 $x^2 - 2 = 7$
13 $y^2 - 1 = 80$
14 $11 - 3x = 2$
15 $4x - 12 = -18$

EXERCISE 12.2H

Solve these equations.

1 $3(x - 2) = 18$
2 $2(1 + x) = 8$
3 $3(x - 5) = 6$
4 $2(x + 3) = 10$
5 $5(x - 2) = 15$
6 $2(x + 3) = 10$
7 $5(x - 4) = 20$
8 $4(x + 1) = 16$
9 $2(x - 7) = 8$
10 $3(2x + 3) = 18$
11 $5(2x - 3) = 15$

12 $2(3x - 2) = 14$
13 $5(2x - 3) = 40$
14 $4(x - 3) = 6$
15 $2(2x - 3) = 8$

EXERCISE 12.3H

Solve these equations.

1 $5x - 1 = 3x + 5$
2 $5x + 1 = 2x + 13$
3 $7x - 2 = 2x + 8$
4 $6x + 1 = 4x + 21$
5 $9x - 10 = 4x + 5$
6 $5x - 8 = 3x - 6$
7 $6x + 2 = 10 - 2x$
8 $2x - 10 = 5 - 3x$
9 $15 + 3x = 2x + 18$
10 $2x - 5 = 4 - x$
11 $3x - 2 = x + 7$
12 $x - 1 = 2x - 6$
13 $2x - 4 = 2 - x$
14 $9 - x = x + 5$
15 $3x - 2 = x - 8$

EXERCISE 12.4H

Solve these equations.

1 $\dfrac{x}{2} = 7$

2 $\dfrac{x}{5} - 2 = 1$

3 $\dfrac{x}{4} + 5 = 8$

4 $\dfrac{x}{3} - 5 = 5$

5 $\dfrac{x}{6} + 3 = 4$

6 $\dfrac{x}{5} + 1 = 4$

7 $\dfrac{x}{8} - 3 = 9$

8 $\dfrac{x}{4} + 1 = 3$

9 $\dfrac{x}{7} + 5 = 6$

10 $\dfrac{x}{4} + 5 = 4$

For each of questions **6** to **15**, solve the inequality.

6 $7 \leqslant 2x - 1$

7 $5x < x + 12$

8 $4x \geqslant x + 9$

9 $4 + x < 0$

10 $3x + 1 \leqslant 2x + 6$

11 $2(x - 3) > x$

12 $5(x + 1) > 3x + 10$

13 $7x + 5 \leqslant 2x + 30$

14 $5x + 2 < 7x - 4$

15 $3(3x + 2) \geqslant 2(x + 10)$

EXERCISE 12.5H

For each of questions **1** to **5**, solve the inequality
and show the solution on a number line.

1 $x - 2 > 1$

2 $x + 1 < 3$

3 $3x - 2 \geqslant 7$

4 $2x + 1 \leqslant 6$

5 $3x - 6 \geqslant 0$

13 → EQUATIONS AND INEQUALITIES 2

EXERCISE 13.1H

Solve each of these equations.

1 $5(2x - 3) = 4x + 3$

2 $3(2x - 1) = 5(2x - 3)$

3 $2(4x - 1) = 6x + 3$

4 $\dfrac{x}{2} = 3x - 15$ **5** $\dfrac{x}{3} = 2x - 5$

6 $\dfrac{3x}{2} = 4 + x$ **7** $\dfrac{5x}{2} = 4x - 3$

8 $\dfrac{2x}{3} = x - \dfrac{1}{2}$ **9** $\dfrac{x}{2} = \dfrac{3x}{4} + \dfrac{1}{2}$

10 $\dfrac{x}{3} = \dfrac{3x}{4} - \dfrac{1}{12}$ **11** $\dfrac{3x}{2} = \dfrac{3x - 2}{5} + \dfrac{11}{5}$

12 $\dfrac{2x - 1}{6} = \dfrac{x - 3}{6} + \dfrac{1}{2}$

13 $\dfrac{x - 2}{3} - \dfrac{2x - 1}{2} = \dfrac{7}{6}$

14 $\dfrac{3x - 2}{2} + \dfrac{2x + 1}{6} + \dfrac{2}{9} = 0$

15 $\dfrac{4x - 3}{2} = \dfrac{x - 3}{3} + \dfrac{7}{6}$

EXERCISE 13.2H

Solve each of these equations.
Where the answer is not exact, give your answer correct to 3 significant figures.

1 $\dfrac{20}{x} = 40$ **2** $\dfrac{48}{x} = 12$ **3** $\dfrac{15}{2x} = 3$

4 $\dfrac{5}{3x} = \dfrac{1}{6}$ **5** $\dfrac{2}{3x} = \dfrac{4}{3}$ **6** $1.4x = 7.6$

7 $3.7x = 40$ **8** $\dfrac{x}{2.3} = 5.6$

9 $7.3(1.2x - 4.7) = 9.6$

10 $\dfrac{1.6}{x} = 2.7$

EXERCISE 13.3H

Solve each of these inequalities.

1 $5(x - 2) > 11 - 2x$

2 $4(3x - 4) \geqslant 3(2x - 1) + 2$

3 $\dfrac{5x}{2} < x + 6$ **4** $\dfrac{2x}{3} + 5 < 4x$

5 $\dfrac{x}{2} > \dfrac{3x}{4} + 2$ **6** $\dfrac{x}{5} \leqslant \dfrac{x}{4} - 2$

7 $3.5x < 4 + x$ **8** $14.6x \geqslant 7.1x + 4.9$

EXERCISE 13.4H

1 Draw a pair of axes and label them 0 to 6 for both x and y.
Show, by shading, the region where $x > 0$, $y > 0$ and $2x + 3y < 12$.

2 Draw a pair of axes and label them -2 to 2 for x and -4 to 8 for y.
Show, by shading, the region where $x < 2$, $y > -2$ and $y < 3x + 2$.

3 Draw a pair of axes and label them 0 to 3 for x and 0 to 9 for y.
Show, by shading, the region where $y < 6$, $y < 3x$ and $y > 2x$.

4 Draw a pair of axes and label them 0 to 8 for both x and y.
Show, by shading, the region where $y > 0$, $y < x$ and $3x + 4y < 24$.

5 Draw a pair of axes and label them 0 to 8 for both x and y.
Show, by shading, the region where $x > 0$, $8x + 3y < 24$ and $5x + 6y > 30$.

EXERCISE 14.1H

1   Draw the graph of $y = 3x$ for values of x from -3 to 3.

2   Draw the graph of $y = x + 2$ for values of x from -4 to 2.

3   Draw the graph of $y = 4x + 2$ for values of x from -3 to 3.

4   Draw the graph of $y = 2x - 5$ for values of x from -1 to 5.

5   Draw the graph of $y = -2x - 4$ for values of x from -4 to 2.

EXERCISE 14.2H

1   Draw the graph of $3y = 2x + 6$ for values of x from -3 to 3.

2   Draw the graph of $2x + 5y = 10$.

3   Draw the graph of $3x + 2y = 15$.

4   Draw the graph of $2y = 5x - 8$ for values of x from -2 to 4.

5   Draw the graph of $3x + 4y = 24$.

EXERCISE 14.3H

1   Simon had a bath. The graph shows the volume (V gallons) of the water in the bath after t minutes.

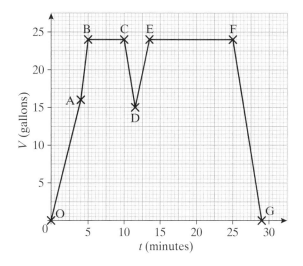

a)   How many gallons of water are in the bath at A?

b)   Simon got in the bath at B and out at F. How long was he in the bath?

c)   Between O and A, the hot tap is on. How many gallons of water per minute came from the hot tap?

d)   Between A and B, both taps are on. What is the rate of flow of both taps together?
    Give your answer in gallons/minute.

e)   Describe what happened between C and E.

f)   At what rate did the bath empty?
    Give your answer in gallons/minute.

2 A printer's charge for printing programmes is worked out as follows.

A fixed charge of £a
+
x pence per programme for the first 1000 programmes
+
80 pence per programme for each programme over 1000

The graph below shows the total charge for printing up to 1000 programmes.

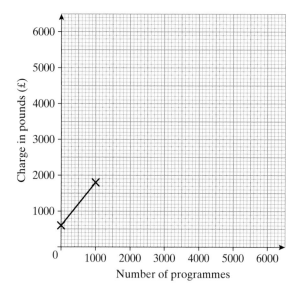

a) What is the fixed charge, £a?
b) Calculate x, the charge per programme for the first 1000 programmes.
c) Copy the graph and add a line segment to show the charges for 1000 to 6000 programmes.
d) What is the total charge for 3500 programmes?
e) What is the average cost per programme for 3500 programmes?

3 The graph shows a train journey.

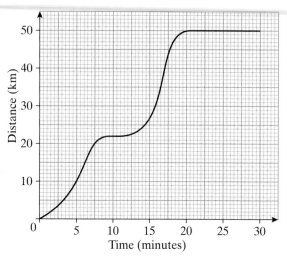

a) How long did the train journey take?
b) How far was the train journey?
c) How far from the start was the first station?
d) How long did the train stop at the first station?
e) When was the train travelling fastest?

4 Water is poured into each of these vessels at a constant rate until they are full.

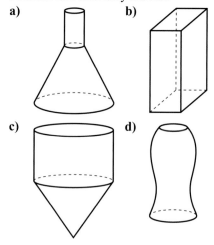

a)   b)

c)   d)

These graphs show depth of water (d) against time (t).
Choose the most suitable graph for each vessel.

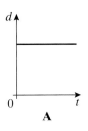

A

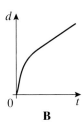

B

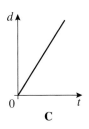

C

D

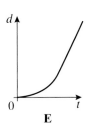

E

F

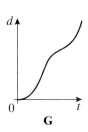

G

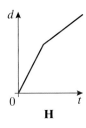

H

5 A mobile phone company offers its customers the choice of two price plans.

	Plan A	Plan B
Monthly subscription	£10.00	£s
Free talk time per month	60 minutes	100 minutes
Cost per minute after the free talk time	a pence	35 pence

The graph shows the charges for Plan A and the charges for Plan B up to 100 minutes.

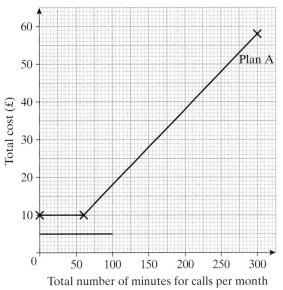

Total number of minutes for calls per month

a) Find the monthly subscription for Plan B (£s).

b) Shamir uses price plan B. How much does it cost him if he uses the phone for 250 minutes per month?

c) Copy the graph and add a line to show the cost for Plan B for 100 to 250 minutes.

d) For how many minutes is the cost the same in both price plans?

e) Which price plan is the cheaper when the time for calls is 220 minutes? By how much?

6 John and Hywel are brothers living in the same house. The graph shows John's cycling trip from their home. He cycles for an hour, stops for a rest, then continues his journey.

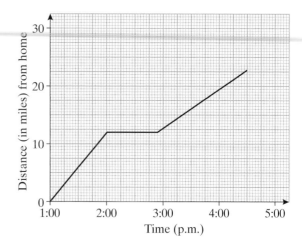

 a) How far did John travel in the first 45 minutes?

 b) For how many minutes did John rest?

 c) Hywel sets out from their home at 3 p.m. and travels by car at an average speed of 25 mph on the same route as John. Draw his journey on a copy of the graph.

 d) Write down the time at which Hywel passes John.

EXERCISE 14.4H

1 Which of these functions are quadratic? For each of the functions that is quadratic, state whether the graph is ∪-shaped or ∩-shaped.

 a) $y = x^2 + 7$ b) $y = 2x^3 + x^2 - 4$ c) $y = x^2 + x - 7$ d) $y = x(6 - x)$

 e) $y = \dfrac{5}{x^2}$ f) $y = x(x^2 + 1)$ g) $y = 2x(x + 2)$ h) $y = 5 + 3x - x^2$

2 a) Copy and complete the table of values for $y = 2x^2$.

x	-3	-2	-1	0	1	2	3
x^2	9					4	
$y = 2x^2$	18					8	

 b) Plot the graph of $y = 2x^2$.
   Use a scale of 2 cm to 1 unit on the x-axis and 1 cm to 1 unit on the y-axis.

 c) Use your graph to
   (i) find the value of y when $x = -1.8$.
   (ii) solve $2x^2 = 12$.

3 a) Copy and complete the table of values for $y = x^2 + x$.

x	-4	-3	-2	-1	0	1	2	3
x^2			4					9
$y = x^2 + x$								12

 b) Plot the graph of $y = x^2 + x$.
   Use a scale of 2 cm to 1 unit on the x-axis and 1 cm to 1 unit on the y-axis.

 c) Use your graph to
   (i) find the value of y when $x = 1.6$.
   (ii) solve $x^2 + x = 8$.

4 a) Copy and complete the table of values for $y = x^2 - x + 2$.

x	-3	-2	-1	0	1	2	3	4
x^2		4						16
$-x$		2						-4
2		2						2
$y = x^2 - x + 2$		8						14

b) Plot the graph of $y = x^2 - x + 2$.
Use a scale of 2 cm to 1 unit on the x-axis and 1 cm to 1 unit on the y-axis.

c) Use your graph to
   (i) find the value of y when $x = 0.7$.
   (ii) solve $x^2 - x + 2 = 6$.

5 a) Copy and complete the table of values for $y = x^2 + 2x - 5$.

x	-5	-4	-3	-2	-1	0	1	2	3
x^2				4					9
$2x$				-4					6
-5				-5					-5
$y = x^2 + 2x - 5$				-5					10

b) Plot the graph of $y = x^2 + 2x - 5$.
Use a scale of 2 cm to 1 unit on the x-axis and 1 cm to 1 unit on the y-axis.

c) Use your graph to
   (i) find the value of y when $x = -1.4$.   **(ii)** solve $x^2 + 2x - 5 = 0$.

6 a) Copy and complete the table of values for $y = 8 - x^2$.

x	-3	-2	-1	0	1	2	3
8				8			8
$-x^2$				0			-9
$y = 8 - x^2$				8			-1

b) Plot the graph of $y = 8 - x^2$.
Use a scale of 2 cm to 1 unit on the x-axis and 1 cm to 1 unit on the y-axis.

c) Use your graph to
   (i) find the value of y when $x = 0.5$.   **(ii)** solve $8 - x^2 = -2$.

7 a) Copy and complete the table of values for $y = (x - 2)(x + 1)$.

x	-3	-2	-1	0	1	2	3	4
$x - 2$		-4					1	
$x + 1$		-1					4	
$y = (x - 2)(x + 1)$		4					4	

b) Plot the graph of $y = (x - 2)(x + 1)$.
Use a scale of 2 cm to 1 unit on the x-axis and 1 cm to 1 unit on the y-axis.

c) Use your graph to
    (i) find the minimum value of y.   **(ii)** solve $(x - 2)(x + 1) = 2.5$.

8 a) Make a table of values for $y = x^2 - 3x + 2$. Choose values of x from -2 to 5.

b) Plot the graph of $y = x^2 - 3x + 2$.
Use a scale of 2 cm to 1 unit on the x-axis and 1 cm to 1 unit on the y-axis.

c) Use your graph to solve
    (i) $x^2 - 3x + 2 = 1$.   **(ii)** $x^2 - 3x + 2 = 10$.

15 ➡ GRAPHS 2

EXERCISE 15.1H

1 Find the gradient of each of these lines.

a)
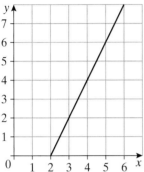

b)
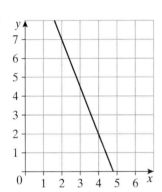

c)
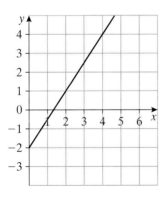

2 Find the gradient of the line joining each of these pairs of points.
a) (1, 2) and (3, 8)
b) (5, 1) and (7, 3)
c) (0, 3) and (2, −3)
d) (−1, 4) and (3, 2)
e) (3, −1) and (−1, −1)

3 Find the gradient of each side of the triangle ABC.

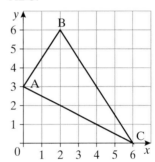

4 Find the gradient of each of these lines.

a)
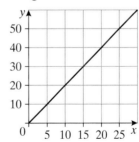

b)
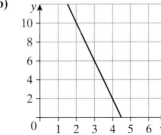

c)

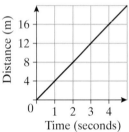

5   Draw the graph of each of these straight lines and find its gradient.
    a) $y = 3x + 1$ **b)** $y = x - 2$
    c) $y = -3x + 2$ **d)** $y = -2x - 1$
    e) $3x + 4y = 12$

6   Find the velocity for these distance–time graphs.
    a)

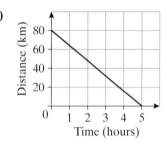

    b)

1   Write down the equations of straight lines with these gradients and y-intercepts.
    a) Gradient 2, y-intercept 6
    b) Gradient −3, y-intercept 5
    c) Gradient 1, y-intercept 0

2   Find the equation of each of these lines.
    a)

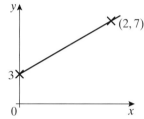

    b)

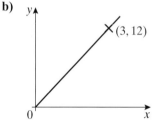

    c)

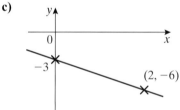

3   Find the gradient and y-intercept of each of these lines.
    a) $y = 4x - 2$ **b)** $y = 3x + 5$
    c) $y = -2x + 1$ **d)** $y = -x + 2$
    e) $y = -3.5x - 8$

4   Find the gradient and y-intercept of each of these lines.
    a) $y + 3x = 7$ **b)** $6x + 2y = 5$
    c) $2x + y = 3$ **d)** $5x - 2y = 8$
    e) $6x + 4y = 9$

5 Find the equation of each of these lines.

a)

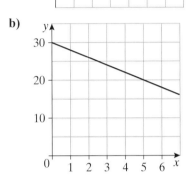

b)

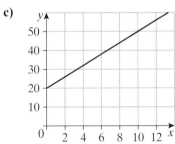

c)

EXERCISE 15.3H

Solve graphically each of these pairs of simultaneous equations.

1 $y = 3x$ and $y = 4x - 2$.
Use values of x from -1 to 4.

2 $y = 2x + 3$ and $y = 4x + 1$.
Use values of x from -2 to 3.

3 $y = x + 4$ and $4x + 3y = 12$.
Use values of x from -3 to 3.

4 $y = 2x + 8$ and $y = -2x$.
Use values of x from -5 to 1.

5 $2y = 3x + 6$ and $3x + 2y = 12$.
Use values of x from 0 to 4.

EXERCISE 15.4H

Use algebra to solve each of these pairs of simultaneous equations.

1 $x + y = 3$
$2x + y = 4$

2 $2x + y = 6$
$2x - y = 2$

3 $2x - y = 7$
$3x + y = 13$

4 $2x + y = 12$
$x - y = 3$

5 $2x + y = 7$
$3x - y = 8$

6 $x + y = 4$
$3x - y = 8$

7 $x + 3y = 9$
$2x - 3y = 0$

8 $3x + y = 14$
$3x + 2y = 22$

9 $4x - y = 4$
$4x + 3y = 20$

10 $x - 4y = 2$
$x + 3y = 9$

EXERCISE 15.5H

Use algebra to solve each of these pairs of simultaneous equations.

1 $x + 3y = 5$
$2x + y = 5$

2 $2x - 5y = 3$
$x + y = 5$

3 $3x - y = 3$
$2x + 3y = 13$

4 $4x - y = 2$
$5x + 3y = 11$

5 $3x - 2y = 8$
$2x - y = 5$

6 $2x + y = 5$
$7x + 2y = 13$

7 $x + 3y = 4$
$3x + 2y = -2$

8 $4x + 3y = 11$
$x + 2y = 4$

9 $x + 2y = 8$
$2x - 3y = 9$

10 $x + y = 2$
$x + 3y = 5$

EXERCISE 16.1H

1 This is a distance–time graph for a particle moving in a straight line.

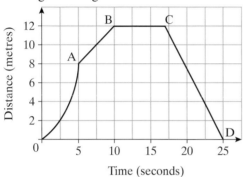

Describe the motion of the particle.

2 **a)** Draw a distance–time graph with the Time axis (*t*) from 0 to 10 seconds and the Distance axis (*s*) from 0 to 20 metres. Draw a straight line joining (0, 0) to (8, 18).
b) Calculate the velocity and describe the motion.

3 **a)** Draw a velocity–time graph with the Time axis (*t*) from 0 to 10 seconds and the Velocity axis (*v*) from 0 to 20 m/s. Show a constant acceleration from $t = 0$, $v = 0$ to $t = 10$, $v = 15$.
b) What is the velocity when $t = 6$?
c) What is the acceleration?

4 A particle moves with a constant acceleration of 2 m/s^2 from $t = 0$ to $t = 3$.
It moves at a constant velocity for the next 5 seconds.
Then it moves with a constant acceleration of -1.25 m/s^2 for the next 12 seconds.
a) Draw a velocity–time graph with the Time axis (*t*) from 0 to 20 seconds and the Velocity axis (*v*) from -10 to $+10$ m/s. Show the movement of the particle.
b) What is the constant velocity from $t = 3$ to $t = 8$?
c) When is the velocity zero?
d) What is the velocity when $t = 20$?

5 This is a velocity–time graph to show the movement of a particle.

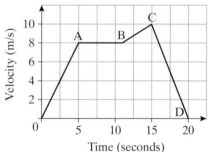

Describe as fully as possible the movement of the particle.
Work out the accelerations where necessary.

6 The graph of $y = x^3 - x^2 - 17x - 15$ is given below.

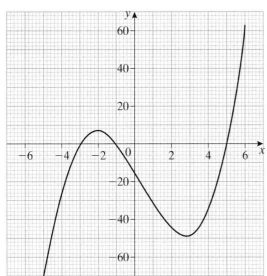

Use the graph of $y = x^3 - x^2 - 17x - 15$ to find
a) the values of x when the gradient of the curve is zero.
b) the gradient of the curve when $x = 2$. (Use tracing paper over the graph.)

EXERCISE 16.2H

1 a) Draw the graph of $y = x^2 - 3x + 2$ for values of x from -1 to 4.
b) On the same axes, draw the line $y = x - 1$.
c) Write down the coordinates of the points where the line and the curve intersect.

2 a) Draw the graph of $y = x^2 - 2x - 3$ for values of x from -2 to 4.
b) On the same axes, draw the line $y = 3 - x$.
c) Write down the coordinates of the points where the line and the curve intersect.

3 a) Draw the graph of $y = x^2 - 6x + 4$ for values of x from 0 to 6.
b) On the same axes, draw the line $4y = 3x - 12$.
c) Write down the coordinates of the points where the line and the curve intersect.

4 a) Draw the graph of $y = 10 + x - 2x^2$ for values of x from -3 to 3.
b) On the same axes, draw the line $3y + 4x = 12$.
c) Write down the coordinates of the points where the line and the curve intersect.

EXERCISE 16.3H

1 a) Draw the graph of $y = x^2 - 5x - 6$ for values of x from -2 to 7.
b) Use your graph to solve these equations.
   (i) $x^2 - 5x - 6 = 0$
   (ii) $x^2 - 5x - 6 = -10$
   (iii) $x^2 - 5x - 6 = 2 - 3x$

2 a) Draw the graph of $y = x^2 - 4x + 3$ for values of x from -1 to 5.
b) Use your graph to solve these equations. Give your answers to 1 decimal place.
   (i) $x^2 - 4x + 3 = 0$
   (ii) $x^2 - 4x - 3 = 0$
   (iii) $x^2 - 6x + 7 = 0$

3 a) Draw the graph of $y = x^2 + 2x - 15$ for values of x from -6 to 4.
b) Use your graph to solve these equations. Give your answers to 1 decimal place.
   (i) $x^2 + 2x - 15 = 0$
   (ii) $x^2 + 2x - 10 = 0$
   (iii) $x^2 + 3x - 4 = 0$
c) (i) What line would you need to draw on the graph to solve the equation $x^2 + x + 3 = 0$?
   (ii) Why does this not work?

For the remaining questions, do not draw the graphs.

4 The graph of $y = x^2 - 2x + 1$ has been drawn. What other line needs to be drawn to solve these equations?
a) $x^2 - 2x + 1 = 5$
b) $x^2 - 2x + 1 = 3x - 2$

5 The graph of $y = x^2 - x - 12$ has been drawn. What other line needs to be drawn to solve each of these equations?
   a) $x^2 - 3x - 12 = 0$
   b) $x^2 - x = 0$
   c) $x^2 + x - 15 = 0$

6 The graph of $y = x^2 - 6x + 8$ has been drawn. What other line needs to be drawn to solve each of these equations?
   a) $x^2 - 6x + 4 = 0$
   b) $x^2 - 8x + 8 = 0$
   c) $x^2 - 4x + 3 = 0$

EXERCISE 16.4H

1 a) Copy and complete this table of values for the equation $y = \dfrac{12}{x}$.

x	1	2	3	4	6	8	12
y							

   b) Draw the graph of $y = \dfrac{12}{x}$ for values of x between 1 and 12.
   c) Use your graph to estimate the value of y when $x = 9$.
   d) (i) Find the point on the graph where $x = y$.
      (ii) Explain the connection of this number with 12.

2 a) Draw the graph of $y = 2x^3$ for values of x between -3 and 3.
   b) Draw the line $y = 10x$ on the same graph.
   c) (i) Find the equation that is given at the intersection of the line and the curve.
      (ii) Estimate the solutions to this equation.

3 a) Draw the graph of $y = \dfrac{6}{x + 1}$ for values of x between 0 and 5.
   b) Use your graph to find the value of x when $y = 4$.
   c) Draw the line $y = x - 1$ on the same graph.
   d) (i) Find the equation that is given at the intersection of the line and the curve.
      (ii) Estimate the solutions to this equation.

4 a) Draw a graph of $y = 4^{-x}$ for values of x between -2 and 2.
   b) Use your graph to estimate
      (i) the value of y when $x = 0.5$.
      (ii) the solution to the equation $4^{-x} = 10$.

5 The table shows two pairs of values for the equation $y = ab^x$.

x	0	1	2	3	4	5
y	4	8				

   a) Find the values of a and b.
   b) Copy and complete the table.
   c) Draw the graph of $y = ab^x$ for values of x from 0 to 5.
   d) Use your graph to estimate the value of x when $y = 80$.

6 The diagram shows the graph of $y = ab^x$. Use the graph to find the values of a and b.

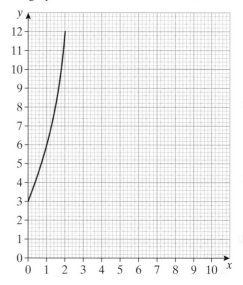

17 → PROPORTION AND VARIATION

EXERCISE 17.1H

1   A carousel revolves 14 times in 252 seconds. How many revolutions does the carousel make in 90 seconds?

2   An express train travels 63 miles in 35 minutes. How long would it take the train to travel 45 miles at the same speed?

3   Sound can travel 5145 metres in 15 seconds. How far can sound travel in 24 seconds?

4   A piece of wax with a volume of 240 cm^3 has a mass of 216 grams. What is the mass of 1000 cm^3 of wax?

5   Tiling a floor with an area of 15 m^2 uses 21 litres of adhesive. How much adhesive would be needed to tile a floor with an area of 35 m^2?

6   A 45 minute telephone call costs £1.80. How much does it cost to make a 12 minute call at the same rate?

7   A contractor is paid £75 for working 6 hours. How much would the contractor be paid for working 10 hours at the same rate?

8   A machine cleans a carpet with an area of 12 m^2 using 21 litres of water. How much water would the machine use to clean a similar carpet with an area of 40 m^2?

9   An aeroplane travels 54 km in 6 minutes. How far would it travel in 15 minutes at the same speed?

10 A person can walk 340 metres in 4 minutes. How far would the same person walk in 3 minutes?

EXERCISE 17.2H

1   A journey takes 30 minutes at a constant speed of 40 miles per hour. How long would the journey take at a constant speed of 60 miles per hour?

2   It takes a team of 12 men 10 weeks to lay a pipeline. How long would the pipeline take to lay if there were 15 men?

3   A pool can be emptied in 18 hours using four pumps. How long will it take to empty the pool using three pumps?

4   A supply of hay is enough to feed eight horses for 30 days. For how long would the same supply feed 20 horses?

5   Using three ploughs it is possible to plough a field in 6 hours. How long would it take to plough the same field using two ploughs?

6   It takes a team of three men 14 hours to install a fence. How long would it take to install the fence if there were eight men?

7   A tank can be filled using three pumps in a period of 28 hours. How long would it take to fill the tank using seven pumps?

8   A carpet with an area of 16 m^2 costs £440. What is the cost of 26 m^2 of the same carpet?

9   A supply of corn is enough to feed 24 pigs for 35 days. For how long would the same supply of corn feed 60 pigs?

10 An express train completes a journey of 473 miles in 5.5 hours. How long would it take the train to travel 129 miles at the same speed?

EXERCISE 17.3H

1 Given that w is directly proportional to f^2, and that $w = 100$ when $f = 5$
   a) find an expression for w in terms of f.
   b) calculate w when $f = 4$.
   c) calculate f when $w = 100$.

2 Given that y is inversely proportional to x^2, and that $y = 8$ when $x = 20$
   a) find an expression for y in terms of x.
   b) calculate
      (i) the value of y when $x = 4$.
      (ii) a value of x when $y = 32$.

3 Given that y is inversely proportional to x, and that $y = 3$ when $x = 10$
   a) find an expression for y in terms of x.
   b) calculate y when $x = 1.5$.
   c) calculate x when $y = 0.5$.

4 The intensity of light, L units, is inversely proportional to the square of the distance, d metres, from the source of light. Given that the intensity of the light 3 metres from the source is 16 units
   a) find an expression for L in terms of d.
   b) calculate the value of d when $L = 36$.

5 Given that y is directly proportional to x^3, and that $y = 32$ when $x = 2$
   a) find an expression for y in terms of x.
   b) calculate:
      (i) the value of y when $x = \frac{1}{3}$.
      (ii) the value of x when $y = 4000$.

EXERCISE 17.4H

For each of these relationships
a) state the type of proportion.
b) find the formula.

1

x	2	6
y	8	72

2

x	1	3
y	36	4

3

x	5	12
y	8.64	1.5

4

x	6	10
y	9	25

5

x	1	5
y	100	4

EXERCISE 17.5H

For each of these relationships
a) state the type of proportion.
b) find the formula.
c) find the missing y value in the table.

1

x	1	8	15
y	6	48	

2

x	4	12	18
y	36	12	

3

x	4	10	36
y	2	12.5	

4

x	3	4	5
y	8	4.5	

5

x	2	5	20
y	30	12	

6

x	4	60	100
y	5	75	

7

x	4	7	40
y	35	20	

8

x	2	3	6
y	9	4	

9

x	1	4	10
y	10	160	

10

x	3	15	75
y	15	75	

11

x	3	4	10
y	24	18	

12

x	2	5	32
y	8	50	

13

x	12	60	150
y	4	20	

14

x	5	10	20
y	8	2	

EXERCISE 18.1H

Find the equation of each of these straight lines.

1

2

3

4

5 A line with gradient -2, passing through the point $(2, 0)$.

6 A line with gradient $\frac{3}{4}$, passing through the point $(4, 2)$.

7 A line passing through $(1, 3)$ and $(4, 9)$.

8 A line passing through $(2, 3)$ and $(5, -6)$.

9 A line passing through $(-1, 5)$ and $(3, -5)$.

10 A line passing through $(3, 1)$ and $(-3, -7)$.

EXERCISE 18.2H

1 Find the gradient of a line perpendicular to the line joining each of these pairs of points.
  a) $(1, 1)$ and $(5, 3)$
  b) $(1, 2)$ and $(4, -2)$
  c) $(-1, 5)$ and $(2, 8)$

2 Find the equation of the line that passes through $(1, 0)$ and is parallel to $y = 2x + 6$.

3 Find the equation of the line that passes through $(2, 3)$ and is parallel to $4x + 2y = 7$.

4 **a)** State the gradient of the line $3x + 5y = 6$.
  b) Find the equation of the line perpendicular to $3x + 5y = 6$ that passes through $(1, 1)$.

5 Find the equation of the line that passes through $(4, 1)$ and is perpendicular to $y = 4x + 3$.

6 Find the equation of the line that passes through $(0, 5)$ and is perpendicular to $3y = x - 1$.

7 Find the equation of the line that passes through (2, 5) and is perpendicular to $7y + 2x = 9$.

8 Which of these lines are
  a) parallel?   **b)** perpendicular?
      $y = x + 5$ $y = 3x + 5$
      $x + 3y = 5$ $4x - y = 5$

9 Two lines cross at right angles at the point (2, 5).
One passes through (4, 7).
What is the equation of the other line?

10 In the diagram AC is a diagonal of the square ABCD. Work out
  a) the equation of the line AC.
  b) the equation of the line BD.
  c) the coordinates of B and D.

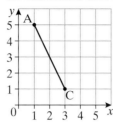

19 → QUADRATIC EQUATIONS

EXERCISE 19.1H

Solve each of these quadratic equations.

1 $x(x + 3) = 0$ **2** $x(x - 4) = 0$

3 $(x - 2)(x + 2) = 0$ **4** $(x - 5)(x + 6) = 0$

5 $3x(x + 5) = 0$ **6** $(x + 10)(x - 1) = 0$

7 $(x - 7)(x - 3) = 0$ **8** $(x + 9)(2x - 5) = 0$

9 $(x - 3)(4x - 1) = 0$ **10** $(3x - 7)(2x + 1) = 0$

11 $4x(3x - 2) = 0$ **12** $(2x - 3)(3x - 4) = 0$

EXERCISE 19.2H

Factorise and hence solve each of these quadratic equations.

1 $x^2 + 5x + 4 = 0$ **2** $x^2 - 8x + 7 = 0$

3 $x^2 + 4x - 5 = 0$ **4** $x^2 - x - 2 = 0$

5 $x^2 + x - 12 = 0$ **6** $x^2 + 7x = 0$

7 $x^2 - 9x + 14 = 0$ **8** $x^2 - 3x - 10 = 0$

9 $3x^2 - 15x = 0$ **10** $x^2 - 16 = 0$

11 $x^2 + 8x + 12 = 0$ **12** $x^2 - 8x - 20 = 0$

13 $x^2 + 8x + 16 = 0$ **14** $x^2 - 8x + 15 = 0$

15 $x^2 - 100 = 0$ **16** $x^2 + 21x - 22 = 0$

17 $2x^2 - 6x = 0$ **18** $x^2 - 11x + 18 = 0$

19 $4x^2 + 2x = 0$ **20** $9x^2 - 25 = 0$

21 $2x^2 + 3x + 1 = 0$ **22** $3x^2 - 5x + 2 = 0$

23 $3x^2 + 10x + 7 = 0$ **24** $2x^2 + 5x - 3 = 0$

25 $3x^2 + 4x - 4 = 0$ **26** $2x^2 - 17x + 8 = 0$

27 $4x^2 - 4x + 1 = 0$ **28** $4x^2 - 11x - 3 = 0$

29 $2x^2 - x - 10 = 0$ **30** $6x^2 - 7x - 5 = 0$

31 $15x^2 + 7x - 2 = 0$ **32** $18x^2 - 35x + 12 = 0$

33 $8x^2 - 16x - 27 = 0$ **34** $12x^2 + 28x - 5 = 0$

35 $28x^2 + 15x + 2 = 0$ **36** $24x^2 - 2x - 15 = 0$

EXERCISE 19.3H

Factorise and hence solve each of these quadratic equations.

1 $x^2 + x = 6$ **2** $x^2 = 7x - 10$

3 $x^2 = 4x + 5$ **4** $x^2 = 10x - 21$

5 $x^2 = 11x$ **6** $x^2 = 12 - 4x$

7 $x^2 = 6 + 5x$ **8** $2x^2 = 3x - 1$

9 $9 + 8x - x^2 = 0$ **10** $5 - 4x - x^2 = 0$

EXERCISE 19.4H

1 **a)** Write each of these quadratic expressions in the form $(x + m)^2 + n$.

   (i) $x^2 + 2x$ **(ii)** $x^2 - 4x$

   (iii) $x^2 - 14x$ **(iv)** $x^2 - x$

   b) Write each of these quadratic expressions in the form $(x + m)^2 + n$.
   Use your answers to part **a)**.

   (i) $x^2 + 2x - 5$ **(ii)** $x^2 - 4x + 7$

   (iii) $x^2 - 14x + 1$ **(iv)** $x^2 - x - 7$

2 For each of these quadratic expressions, complete the square; that is, write it in the form $(x + m)^2 + n$.

   a) $x^2 + 6x - 1$ **b)** $x^2 + 8x - 2$

   c) $x^2 - 4x + 3$ **d)** $x^2 - 2x - 3$

   e) $x^2 - 12x + 37$ **f)** $x^2 + 10x - 3$

   g) $x^2 - 6x + 19$ **h)** $x^2 + 3x - 2$

   i) $x^2 - 5x + 7$

EXERCISE 19.5H

In this exercise, give all your answers correct to 2 decimal places.

1 Solve each of these quadratic equations.

   a) $(x - 2)^2 - 5 = 0$

   b) $(x + 3)^2 - 7 = 0$

   c) $(x - 4)^2 - 20 = 0$

   d) $(x + 1)^2 - 11 = 0$

2 For each of these quadratic equations, first complete the square and then solve the equation.

a) $x^2 - 6x - 5 = 0$
b) $x^2 + 4x + 1 = 0$
c) $x^2 - 4x - 3 = 0$
d) $x^2 + 10x + 5 = 0$
e) $x^2 + 2x - 8 = 0$
f) $x^2 + 6x - 2 = 0$
g) $x^2 - 8x + 6 = 0$
h) $x^2 + 14x - 3 = 0$
i) $x^2 + 3x - 6 = 0$

EXERCISE 19.6H

Solve each of these quadratic equations using the formula.
Give your answers correct to 2 decimal places.
If there are no real solutions, say so.

1 $x^2 + 3x + 1 = 0$ 2 $x^2 - 7x + 3 = 0$

3 $x^2 + 2x - 11 = 0$ 4 $x^2 + x - 7 = 0$

5 $2x^2 + 6x + 3 = 0$ 6 $2x^2 + 5x + 1 = 0$

7 $x^2 + 4x + 7 = 0$ 8 $3x^2 + 10x + 5 = 0$

9 $3x^2 - 7x - 4 = 0$ 10 $2x^2 - x - 8 = 0$

11 $5x^2 + 8x + 1 = 0$ 12 $2x^2 + 5x - 7 = 0$

13 $4x^2 - 12x + 6 = 0$ 14 $2x^2 - 5x - 4 = 0$

15 $2x^2 + 10x + 15 = 0$ 16 $3x^2 + 3x - 2 = 0$

17 $x^2 - 3x - 50 = 0$ 18 $4x^2 + 9x + 3 = 0$

19 $5x^2 + 11x + 4 = 0$ 20 $3x^2 - x - 8 = 0$

21 $2x^2 + 6x + 5 = 0$

EXERCISE 19.7H

1 A solid cuboid has length $(x + 3)$ cm, width $(x + 3)$ cm and height 5 cm. The surface area of the cuboid is 195 cm².

a) Show that x satisfies the equation
$2x^2 + 32x - 117 = 0$.
b) Use the formula method to solve the equation $2x^2 + 32x - 117 = 0$, giving your answers correct to 2 decimal places. Show your working.
c) Hence write down the dimensions of the cuboid.

2 For the first x seconds of a journey the average speed of a cyclist is 4 m/s. For the next $(5x + 2)$ seconds, the average speed is x m/s. The total distance travelled is 128 metres.

a) Show that x satisfies the equation
$5x^2 + 6x - 128 = 0$.
b) Use the formula method to solve the equation $5x^2 + 6x - 128 = 0$, giving solutions correct to 1 decimal place.
c) Hence find the total time for the journey.

3 The surface area of a cuboid with height x cm, length $(x + 2)$ cm and width x cm is 83 cm².

a) Show that x satisfies the equation
$6x^2 + 8x - 83 = 0$.
b) Use the formula method to solve the equation $6x^2 + 8x - 83 = 0$, giving your answers correct to 2 decimal places.
c) Hence write down the dimensions of the cuboid.

4 The diagram shows a triangular prism.

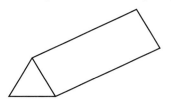

The area of a cross-section of the triangular prism is $2x^2$ cm² and the area of each of its rectangular faces is $(7x + 5)$ cm². The surface area of the triangular prism is 202 cm².

a) Show that x satisfies the equation
$4x^2 + 21x - 187 = 0$.
b) Use the formula method to solve the equation $4x^2 + 21x - 187 = 0$, giving solutions correct to 1 decimal place.
c) Hence find the area of the cross-section of this triangular prism and the area of each rectangular face.

EXERCISE 20.1H

Solve each of these pairs of simultaneous equations.

1 $2x + y = 8$
   $3x - y = 7$

2 $4x + 5y = 3$
   $3x + 5y = 1$

3 $2x + 5y = 6$
   $2x + 3y = 14$

4 $4x - 5y = 7$
   $4x + y = -11$

5 $2x + 3y = 9$
   $6x - 3y = 39$

6 $3x + 5y = 23$
   $7x + 5y = 37$

7 $5x + 2y = 5$
   $-5x - 3y = 5$

8 $x + 3y = 13$
   $2x + y = 11$

9 $3x + 2y = 16$
   $5x + 4y = 27$

10 $4x - 3y = -3$
   $2x + 5y = 18$

11 $3x - 2y = 19$
   $x + 4y = -3$

12 $4x - 3y = 1$
   $x + y = -5$

13 $3x + 2y = 4$
   $x + 4y = 13$

14 $5x + y = 10$
   $3x - 5y = 10$

15 $7x + 3y = 0$
   $2x - 9y = 69$

16 $3x + 4y = 12$
   $6x - 3y = 2$

17 $4x - 3y = 11$
   $5x - y = 22$

18 $3x + 4y = 19$
   $5x + 3y = 17$

19 $3x - 2y = 13$
   $7x + 6y = 9$

20 $6x - 5y = -16$
   $5x - 3y = -11$

21 $4x + 2y = 17$
   $3x + 5y = 18$

EXERCISE 20.2H

Solve each of these pairs of simultaneous equations by substitution.

1 $y = 2x - 10$
   $y = 3x - 13$

2 $y = 5x + 18$
   $y = 4 - 2x$

3 $y = 3x + 11$
   $x + y = 3$

4 $y = 8 - 2x$
   $2x + 5y = 48$

5 $x = 3y - 5$
   $3x - 2y = 6$

6 $x - 2y = 7$
   $y = 3x + 4$

21 → SEQUENCES

EXERCISE 21.1H

1 Look at this sequence of circles.
The first four patterns in the sequence have been drawn.

  a) How many circles are there in the 100th pattern?
  b) Describe the position-to-term rule for this sequence.

2 Look at this sequence of matchstick patterns.

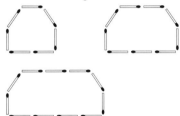

  a) Copy and complete this table.

Pattern number	1	2	3	4	5
Number of matchsticks					

  b) What patterns can you see in the numbers?
  c) Find the number of matchsticks in the 50th pattern.

3 Here is a sequence of star patterns.

  a) Draw the next pattern in the sequence.
  b) Without drawing the pattern, find the number of stars in the 8th pattern.
    Explain how you found your answer.

4 The numbers in a sequence are given by the rule multiply the position number by 7, then subtract 10.

  a) Show that the first term of the sequence is -3.
  b) Find the next four terms in the sequence.

5 Find the first four terms of the sequences with these nth terms.
  a) $10n$ **b)** $8n + 2$

6 Find the first five terms of the sequences with these nth terms.
  a) n^2 **b)** $2n^2$ **c)** $5n^2$

7 The first term of a sequence is 3.
The general rule for the sequence is multiply a term by 3 to get to the next term.
Write down the first five terms of the sequence.

8 For a sequence, $T_1 = 12$ and $T_{n+1} = T_n - 5$.
Write down the first four terms of this sequence.

9 Draw suitable patterns to represent this sequence.

  $$1, 4, 7, 10, ...$$

10 Draw suitable patterns to represent this sequence.

$$1 \times 1, 3 \times 3, 5 \times 5, 7 \times 7, ...$$

EXERCISE 21.2H

1 Find the nth term for each of these sequences.
  a) 10, 13, 16, 19, 22, ...
  b) 0, 1, 2, 3, 4, ...
  c) −3, −1, 1, 3, 5, ...

2 Find the nth term for each of these sequences.
  a) 25, 20, 15, 10, 5, ...
  b) 4, 2, 0, −2, −4, ...
  c) 3, 2, 1, 0, −1, ...

3 Which of these sequences are linear?
  Find the next two terms of each of the sequences that are linear.
  a) 2, 5, 10, 17, ...
  b) 2, 5, 8, 11, ...
  c) 1, 3, 6, 10, ...
  d) 12, 8, 4, 0, −4, ...

4 **a)** Write down the first five terms of the sequence with nth term $100n$.
  b) Compare your answers with this sequence.
    99, 199, 299, 399, ...
    Write down the nth term of this sequence.

5 A mail-order shirt company charges £25 per shirt, plus an overall delivery charge of £3.
  a) Copy and complete the table.

Number of shirts	1	2	3
Cost in £			

  b) Write an expression for the cost, in pounds, of n shirts.
  c) Paul pays £128 for shirts. How many does he buy?

6 **a)** Write down the first five terms of the sequence with nth term n^2.
  b) Compare your answers with this sequence.
    0, 3, 8, 15, 24, ...
    Write the nth term of this sequence.

7 The nth triangle number is $\dfrac{n(n + 1)}{2}$.
  Find the 60th triangle number.

8 The nth term of a sequence is 2^n.
  a) Write down the first five terms of this sequence.
  b) Describe the sequence.

9 **a)** Write down the first five cube numbers.
  b) Compare the sequence below with the sequence of cube numbers.
    3, 10, 29, 66, 127, ...
    Use what you notice to write down the nth term of this sequence.
  c) Find the 10th term of this sequence.

10 **a)** Compare the sequence below with the sequence of square numbers.
    5, 20, 45, 80, 125, ...
    Use what you notice to write down the nth term of this sequence.
  b) Find the 20th term of this sequence.

11 Find the nth term of each of the following sequences.
  a) $1 \times 2, 2 \times 3, 3 \times 4, 4 \times 5, ...$
  b) $1 \times 3, 2 \times 5, 3 \times 7, 4 \times 9, ...$
  c) $2 \times 1, 4 \times 3, 6 \times 5, 8 \times 7, ...$

EXERCISE 22.1H

1 Pearl is a child minder. She charges £6.50 an hour.
She looks after Mrs Khan's child for 6 hours. How much does she charge?

2 It costs £60 plus £1 a mile to hire a coach.
 a) How much does it cost to hire a coach to go
  (i) 80 miles? (ii) 150 miles?
 b) Write a formula for the cost, £C, of hiring a coach to go n miles.

3 The cost of booking a room for a meeting is £80 plus £20 an hour.
 a) How much does it cost to hire the room for
  (i) 5 hours? (ii) 8 hours?
 b) Write a formula for the cost, £C, of hiring the room for h hours.

4 The perimeter of a rectangle is twice the length plus twice the width.
 a) What is the perimeter of a rectangle with length 5 cm and width 3.5 cm?
 b) Write a formula for the perimeter, P, of a rectangle with length x and width y.

5 To find the volume of a pyramid, multiply the area of the base by the height and divide by 3.
 a) What is the volume of a pyramid with base area 12 cm^2 and height 7 cm?
 b) Write a formula for the volume, V, of a pyramid with base area A and height h.

6 To find the time it takes to type a document, divide the number of words in the document by the number of words typed per minute.
 a) How long does it take Liz to type a document 560 words long if she types 80 words a minute?
 b) Write a formula for the time, T, to type a document w words long if the typist types r words a minute.

7 The time, t, for a journey is the distance, d, divided by the speed, s.
 a) Write a formula for this.
 b) Steve travelled 175 miles at a speed of 50 mph. How long did the journey take?

8 The circumference of a circle is given by the formula:

   $C = \pi \times D$, where D is the diameter of the circle.

 Find the circumference of a circle with diameter 8.5 cm. Use $\pi = 3.14$.

9 At Carterknowle toddlers group, the charge is £1 per carer and 50p for every toddler they bring.
 a) Tracey brings three toddlers to the group. How much does she pay?
 b) Fran brings n toddlers to the group. Write down an equation for the amount, £A, she has to pay.

10 a) For the formula $A = b - c$, find A when $b = 6$ and $c = 3.5$.
 b) For the formula $B = 2a - b$, find B when $a = 6$ and $b = 5$.
 c) For the formula $C = 2a - b + 3c$, find C when $a = 3.5$, $b = 2.6$ and $c = 1.2$.
 d) For the formula $D = 3b^2$, find D when $b = 2$.
 e) For the formula $E = ab - cd$, find E when $a = 12.5$, $b = 6$, $c = 3.5$ and $d = 8$.
 f) For the formula $F = \dfrac{a - b}{5}$, find F when $a = 6$ and $b = 3.5$.

EXERCISE 22.2H

1   Rearrange each of these formulae to make the letter in brackets the subject.
   a) $a = b + c$ (b)
   b) $a = 3x - y$ (x)
   c) $a = b + ct$ (t)
   d) $F = 2(q + p)$ (q)
   e) $x = 2y - 3z$ (y)
   f) $P = \dfrac{3 + 4n}{5}$ (n)

2   The formula for the circumference of a circle is $C = \pi d$.
    Rearrange the formula to make d the subject.

3   Rearrange the formula $A = \dfrac{3ab}{2n}$ to make
   a) a the subject.
   b) n the subject.

4   The formula for finding the perimeter of a rectangle is $P = 2(a + b)$, where P is the perimeter, a is the length of the rectangle and b is the width of the rectangle.
    Rearrange the formula to make a the subject.

5   The formula $y = mx + c$ is the equation of a straight line.
    Rearrange it to find m in terms of x, y and c.

6   The surface area of a sphere is given by the formula $A = 4\pi r^2$.
    Rearrange the formula to make r the subject.

7   The formula for the volume of this prism is $V = \dfrac{\pi r^2 h}{4}$.

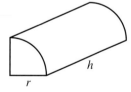

   a)   Find V when $r = 2.5$ and $h = 7$.
   b)   (i)   Rearrange the formula to make r the subject.
        (ii)   Find r when $V = 100$ and $h = 10$.

8   The formula for the surface area of a closed cylinder is $S = 2\pi r(r + h)$.
    Rearrange the formula to make h the subject.

EXERCISE 22.3H

1   Calculate the value of $x^3 + x$ when
   a) $x = 2$.   b) $x = 3$.   c) $x = 2.5$.

2   a)   Calculate the value of $x^3 - x$ when
        (i) $x = 4$.   (ii) $x = 5$.
        (iii) $x = 4.6$.   (iv) $x = 4.7$.
        (v) $x = 4.65$.
   b)   Using your answers to part a), give the solution of $x^3 - x = 94$, to 1 decimal place.

Use trial and improvement for questions 3 to 10. Show your trials.

3   Find a solution, between $x = 2$ and $x = 3$, to the equation $x^3 = 11$.
    Give your answer correct to 1 decimal place.

4   a)   Show that a solution to the equation $x^3 + 3x = 30$ lies between $x = 2$ and $x = 3$.
   b)   Find the solution correct to 1 decimal place.

5   a)   Show that a solution to the equation $x^3 - 2x = 70$ lies between $x = 4$ and $x = 5$.
   b)   Find the solution correct to 1 decimal place.

6   Find a solution to the equation $x^3 + 4x = 100$.
    Give your answer correct to 1 decimal place.

7   Find a solution to the equation $x^3 + x = 60$.
    Give your answer correct to 2 decimal places.

8   Find a solution to the equation $x^3 - x^2 = 40$.
    Give your answer correct to 2 decimal places.

9   A number, x, added to the square of that number, is equal to 1000.
   a)   Write this as a formula.
   b)   Find the number correct to 1 decimal place.

10   The cube of a number minus the number is equal to 600.
    Find the number correct to 2 decimal places.

23 → FORMULAE 2

EXERCISE 23.1H

Rearrange each of these formulae to make the letter in brackets the subject.

1 $C = 2\pi r$ (r)

2 $k = \dfrac{PV}{T}$ (V)

3 $ax + b = 2x + 3b$ (x)

4 $a(x - y) = 3(x + y)$ (x)

5 $pq - r = rq - t$ (p)

6 $pq - r = rq - t$ (q)

7 $pq - r = rq - t$ (r)

8 $s = \frac{1}{2}(u + v)t$ (t)

9 $s = \frac{1}{2}(u + v)t$ (v)

10 $V = \dfrac{1}{x} - \dfrac{1}{3}$ (x)

11 $P = \dfrac{3g + f}{4g - e}$ (g)

EXERCISE 23.2H

1 Rearrange each of these formulae to make the letter in brackets the subject.

a) $y = 2x^2 + 3$ (x)

b) $h = \dfrac{gt^2}{4\pi^2}$ (t)

c) $y = \sqrt{\dfrac{x}{3}}$ (x)

d) $y = \sqrt{x + a}$ (x)

e) $s = \sqrt{x^2 + y^2}$ (x)

f) $A = \frac{1}{3}\pi r\,(l + 3r)$ (l)

g) $y = \dfrac{2x}{3} - 5$ (x)

h) $y = (x - a)^2$ (x)

2 The formula for the volume of a sphere is $V = \frac{4}{3}\pi r^3$ where r is the radius of the sphere.

a) Find the volume of a sphere of radius 3 cm.
    Give your answer to 1 decimal place.

b) Rearrange the formula to make r the subject.

c) What is the radius of a sphere with volume 500 cm³?
    Give your answer to 1 decimal place.

3 The formula for the geometric mean, m, of two numbers a and b is $m = \sqrt{ab}$.
    For three numbers, a, b and c, the geometric mean is $m = \sqrt[3]{abc}$, and for four numbers, a, b, c and d, it is $\sqrt[4]{abcd}$.

a) Use the appropriate formula to find the geometric mean of the numbers 5, 8 and 12.
    Give your answer correct to 2 decimal places.

b) The geometric mean of four numbers 4, 7, 8 and x is 6.88, correct to 2 decimal places.
    (i) Rearrange the appropriate formula to make x the subject.
    (ii) Find x correct to the nearest whole number.

EXERCISE 24.1H

1   $f(x) = x^2 + 4$.
    Find the value of these.
    a) $f(3)$ **b)** $f(-2)$

2   $g(x) = x^2 + 2x + 1$.
    Find the value of these.
    a) $g(3)$ **b)** $g(-2)$ **c)** $g(0)$

3   $h(x) = 5x - 3$
    a) Solve $h(x) = 7$.
    b) Write an expression for these.
       (i) $h(x - 2)$ **(ii)** $h(2x)$

4   $f(x) = 2x + 4$
    a) Solve $f(x) = 1$.
    b) Write an expression for these.
       (i) $2f(x)$ **(ii)** $f(2x + 3)$

5   $g(x) = 5x - 3$
    a) Solve $g(x) = 0$.
    b) Write an expression for these.
       (i) $g(x + 3)$ **(ii)** $2g(x) + 3$

6   $h(x) = x^2 + 2$
    a) Solve $h(x) = 6$.
    b) Write an expression for these.
       (i) $h(x + 3)$ **(ii)** $h(2x) + 1$

7   $f(x) = 2x^2 - 3x$
    a) Find the value of $f(-2)$.
    b) Write an expression for these.
       (i) $f(x + 2)$ **(ii)** $f(3x)$

8   $g(x) = x^2 + 3x$
    a) Solve $g(x) = -2$.
    b) Write an expression for these.
       (i) $3g(x) + 4$ **(ii)** $g(2x + 1)$

EXERCISE 24.2H

1   **a)** Sketch these graphs on the same diagram.
       (i) $y = -x^2$
       (ii) $y = 2 - x^2$
    b) State the transformation that maps
       $y = -x^2$ on to $y = 2 - x^2$.

2   **a)** Sketch these graphs on the same diagram.
       (i) $y = x^2$
       (ii) $y = x^2 - 4$
    b) State the transformation that maps $y = x^2$
       on to $y = x^2 - 4$.

3   **a)** Sketch these graphs on the same diagram.
       (i) $y = x^2$
       (ii) $y = (x - 2)^2$
       (iii) $y = (x - 2)^2 + 3$
    b) State the transformation that maps $y = x^2$
       on to $y = (x - 2)^2 + 3$.

4   **a)** Sketch the result of translating the graph
       of $y = \cos \theta$ by $\begin{pmatrix} 0 \\ 2 \end{pmatrix}$.
    b) State the equation of the transformed
       graph.

5   State the equation of $y = \tan \theta$ after it has
    been translated by these vectors.
    a) $\begin{pmatrix} 0 \\ 4 \end{pmatrix}$ **b)** $\begin{pmatrix} 3 \\ 0 \end{pmatrix}$

6 The diagram shows the graph of $y = f(x)$.

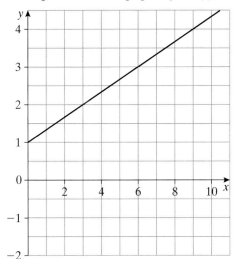

Copy the diagram and sketch these graphs on the same axes.

a) $y = f(x) - 3$ **b)** $y = f(x - 3)$

7 The diagram shows the graph of $y = g(x)$.

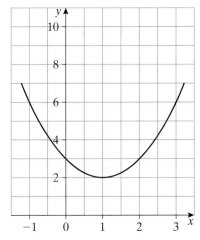

Copy the diagram and sketch these graphs on the same axes.

a) $y = g(x - 1)$ **b)** $y = g(x) - 2$

8 State the equation of the graph $y = x^2$ after it has been translated by $\begin{pmatrix} 3 \\ -4 \end{pmatrix}$.

9 This is the graph of a transformed sine curve. State its equation.

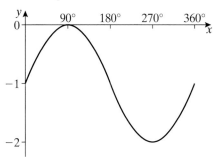

10 The graph of $y = x^2 + 3x$ is translated by $\begin{pmatrix} 2 \\ 3 \end{pmatrix}$.

a) State the equation of the transformed graph.

b) Show that this equation can be written as $y = x^2 - x + 1$.

EXERCISE 24.3H

1 a) The diagram shows a sketch of $y = f(x)$. Copy the sketch. On the same diagram, sketch the curve $y = f(x) + 4$.
Mark clearly the coordinates of the point where the curve crosses the y-axis.

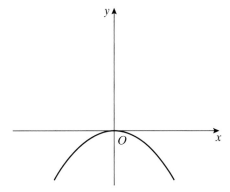

b) The diagram shows a sketch of $y = g(x)$.
Copy the sketch. On the same diagram, sketch the curve $y = -g(x)$.

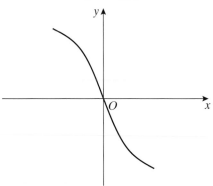

2 **a)** The diagram shows a sketch of $y = f(x)$.
Copy the sketch. On the same diagram, sketch the curve $y = f(x + 6)$.
Mark clearly the coordinates of the point where the curve touches the x-axis.

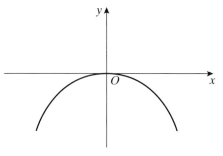

c) The diagram shows a sketch of $y = h(x)$.
Copy the sketch. On the same diagram, sketch the curve $y = h(4x)$.

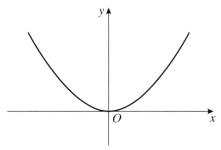

b) The diagram shows a sketch of $y = g(x)$.
Copy the sketch. On the same diagram, sketch the curve $y = g(x) + 6$.
Mark clearly the coordinates of the point where the curve crosses the y-axis.

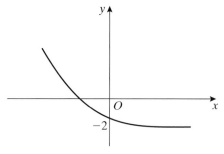

d) The diagram shows a sketch of $y = j(x)$.
Copy the sketch. On the same diagram, sketch the curve $y = j(x - 2)$.
Mark clearly the coordinates of the point where the curve crosses the x-axis.

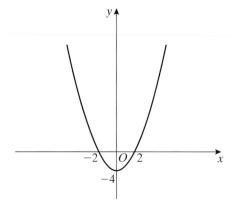

c) The diagram shows a sketch of $y = x^2$.
Copy the sketch. On the same diagram, sketch the curves
i) $y = -2x^2$
ii) $y = 3 - 2x^2$

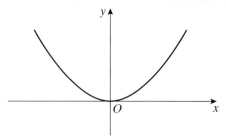

3 a) The diagram shows a sketch of the curve $y = x^2$. Copy the sketch and on the same diagram, sketch the curve $y = 3x^2$.

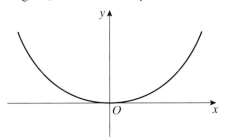

b) The diagram shows a sketch of the curve $y = g(x)$. Copy the sketch and on the same diagram, sketch the curve $y = g(x) - 3$. Mark clearly the coordinates of the point where the curve crosses the y-axis.

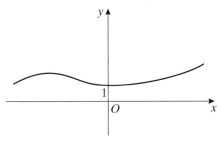

c) The diagram shows a sketch of the curve $y = h(x)$. Copy the sketch and on the same diagram, sketch the curve $y = h(x - 2)$.

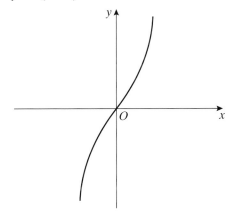

EXERCISE 24.4H

1 a) Sketch on the same axes the graphs of $y = \cos \theta$ and $y = 2 \cos \theta$ for $0° \leqslant \theta \leqslant 360°$.

b) Describe the transformation that maps $y = \cos \theta$ on to $y = 2 \cos \theta$.

2 a) Sketch on the same axes the graphs of $y = \sin \theta$ and $y = \sin 2\theta$ for $0° \leqslant \theta \leqslant 360°$.

b) Describe the transformation that maps $y = \sin \theta$ on to $y = \sin 2\theta$.

3 Describe the transformation that maps
a) $y = \sin \theta + 1$ on to $y = \sin(-\theta) + 1$.
b) $y = x^2 + 2$ on to $y = -x^2 - 2$.
c) $y = x^2$ on to $y = 3x^2$.
d) $y = \sin \theta$ on to $y = \sin \dfrac{\theta}{3}$.

4 The graph of $y = \sin \theta$ is transformed by a one-way stretch parallel to the θ-axis with a scale factor of $\frac{1}{4}$.
State the equation of the resulting graph.

5 State the equation of the graph $y = x^2 - 1$ after each of these transformations.
a) A reflection in the y-axis
b) A reflection in the x-axis

6 State the equation of the graph of $y = 4x + 1$ after each of these transformations.
a) A one-way stretch parallel to the y-axis with a scale factor of 4
b) A one-way stretch parallel to the x-axis with a scale factor of $\frac{1}{2}$

7 Describe the transformation that maps $y = f(x)$ on to each of these graphs.
a) $y = f(x) - 2$ **b)** $y = 3f(x)$
c) $y = f(0.5x)$ **d)** $y = 4f(2x)$

8 State the equation of the graph $y = x^2 + 3$ after each of these transformations.
a) A reflection in the x-axis
b) A reflection in the y-axis
c) A one-way stretch parallel to the x-axis with a scale factor of 0.5

9 The graph of $y = x^2 - 2x$ is stretched parallel
   to the x-axis by a scale factor of 2.
   a) State the equation of the resulting graph.
   b) What does point $(1, -1)$ map on to under
       this transformation?

10 The equation of this graph is $y = a \sin b\theta$.
    Find a and b.

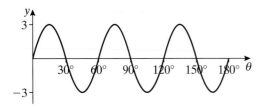

EXERCISE 25.1H

Simplify these.

1. $\dfrac{x+4}{3} + \dfrac{x-1}{2}$

2. $\dfrac{x+5}{3} - \dfrac{x+3}{4}$

3. $\dfrac{x-3}{2} - \dfrac{x-5}{3}$

4. $\dfrac{4x+3}{2} - \dfrac{3x-2}{4}$

5. $\dfrac{2}{x+3} + \dfrac{x-1}{3}$

6. $\dfrac{2}{x-5} + \dfrac{x+4}{2}$

7. $\dfrac{3x+4}{5} - \dfrac{3}{2x-1}$

8. $\dfrac{5x+4}{3x-2} + \dfrac{5}{4}$

EXERCISE 25.2H

Simplify these.

1. $\dfrac{3}{x+2} + \dfrac{2}{x-1}$

2. $\dfrac{4}{x-3} + \dfrac{x+3}{2x}$

3. $\dfrac{x+5}{2x} - \dfrac{3x}{x+2}$

4. $\dfrac{x+5}{x-3} - \dfrac{x-3}{x+2}$

5. $\dfrac{1}{x+4} - \dfrac{3}{x-5}$

6. $\dfrac{x}{x-3} + \dfrac{x-2}{x+1}$

7. $\dfrac{3x+2}{x-5} - \dfrac{x-3}{x-4}$

8. $\dfrac{2x+3}{3x+1} - \dfrac{x-2}{2x+5}$

EXERCISE 25.3H

Solve these.

1. $\dfrac{x-3}{2} - \dfrac{x-2}{3} = 1$

2. $\dfrac{5}{x-4} = \dfrac{3}{x-2}$

3. $\dfrac{1}{4x-3} = \dfrac{1}{3x+2}$

4. $\dfrac{3-x}{4} + \dfrac{2x+5}{3} = 1$

5. $\dfrac{x-2}{5} - \dfrac{2x-3}{4} = \dfrac{1}{3}$

6. $\dfrac{3x}{x+4} + \dfrac{2x}{5x-2} = \dfrac{3}{2}$

7. $\dfrac{2}{3x+1} - \dfrac{5}{x+3} = 0$

8. $\dfrac{3}{x-2} - \dfrac{1}{x+1} = 1$

9. $\dfrac{2}{2x+3} + \dfrac{1}{x+2} = 3$

EXERCISE 26.1H

Find the area of each of these triangles.

1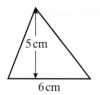
5 cm
6 cm

2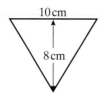
10 cm
8 cm

3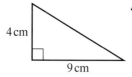
4 cm
9 cm

4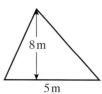
8 m
5 m

5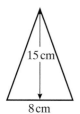
15 cm
8 cm

6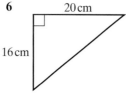
20 cm
16 cm

7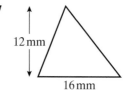
12 mm
16 mm

8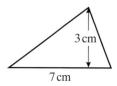
3 cm
7 cm

9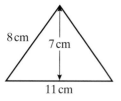
8 cm
7 cm
11 cm

10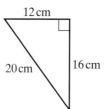
12 cm
20 cm
16 cm

EXERCISE 26.2H

Find the area of each of these parallelograms.

1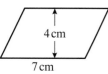
4 cm
7 cm

2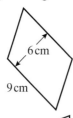
6 cm
9 cm

3
10 cm
2.5 cm

4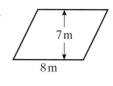
7 m
8 m

5
5 cm 6 cm
4 cm

6
4 cm
5.5 cm

7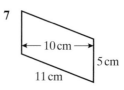
10 cm
5 cm
11 cm

8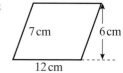
7 cm 6 cm
12 cm

Find the missing length in each of these diagrams.

9

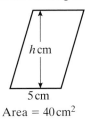

h cm

5 cm

Area = 40 cm^2

10

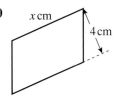

x cm

4 cm

Area = 36 cm^2

EXERCISE 26.3H

Find the size of the lettered angles. Give a reason for each answer.

1

a

60°

2

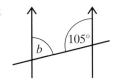

105°

b

3

65°

c

4

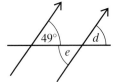

49°

d

e

5

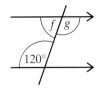

f g

120°

6

100°

h i

7

k

j

60°

8

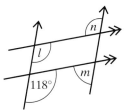

n

l

m

118°

9

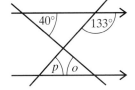

40° 133°

p o

10

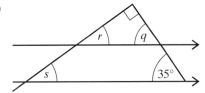

r q

s 35°

EXERCISE 26.4H

Find the size of the lettered angles. Give a reason for each answer.

1

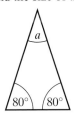

a

80° 80°

2

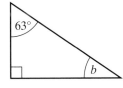

63°

b

3

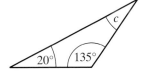

c

20° 135°

4

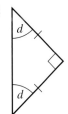

d

d

5

38°

e   117°

6

10°

150°

f

7

80°

52°   h   g

8

70°   i

105°   110°

9

92°

83°

k j   70°

10

55°

m

54°   l 65°

EXERCISE 26.5H

1   Name each of these quadrilaterals.

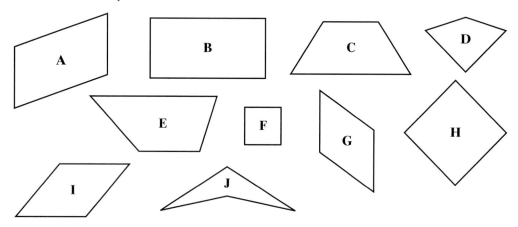

A

B

C

D

E

F

G

H

I

J

2   Name the quadrilateral or quadrilaterals which have the following properties.
   a)   Four right angles
   b)   Both pairs of opposite sides parallel
   c)   Equal diagonals
   d)   At least one pair of opposite sides parallel
   e)   Diagonals that bisect each other

3   Plot each set of points on squared paper and join them in order to make quadrilateral. Use a different grid for each part.
   Write down the special name of each quadrilateral.
   a)   (3, 0), (5, 2), (3, 4), (1, 2)
   b)   (2, 1), (4, 1), (4, 5), (2, 5)
   c)   (1, 2), (3, 1), (3, 6), (1, 7)
   d)   (2, 1), (2, 5), (8, 4), (8, 2)

4 A rhombus is a special type of parallelogram.
What extra properties does a rhombus have?

5 A quadrilateral has angles of 80°, 100°, 80°, 100° in order as you work your way around the quadrilateral. The sides are not all the same length.
Which special quadrilateral could have these as its angles?
Draw the quadrilateral and mark on the angles.

EXERCISE 26.6H

1 A polygon has nine sides.
Work out the sum of the interior angles of this polygon.

2 A polygon has 13 sides.
Work out the sum of the interior angles of this polygon.

3 Four of the exterior angles of a hexagon are 93°, 50°, 37° and 85°.
The other two angles are equal.
   a) Work out the size of these equal exterior angles.
   b) Work out the size of the interior angles of the hexagon.

4 Four of the interior angles of a pentagon are 170°, 80°, 157° and 75°.
   a) Work out the size of the other interior angle.
   b) Work out the size of the exterior angles of the pentagon.

5 A regular polygon has 18 sides.
Find the size of the exterior and the interior angles of this polygon.

6 A regular polygon has 24 sides.
Find the size of the exterior and the interior angles of this polygon.

7 A regular polygon has an exterior angle of 12°.
Work out the number of sides that the polygon has.

8 A regular polygon has an interior angle of 172°.
Work out the number of sides that the polygon has.

EXERCISE 27.1H

For each of these diagrams, find the area of the third square.

1

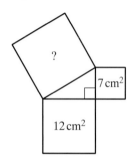

2

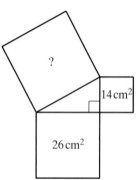

3

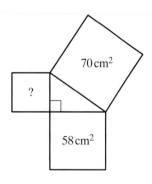

4

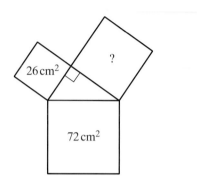

EXERCISE 27.2H

1 For each of these triangles, find the length marked x.
Where the answer is not exact, give your answer correct to 2 decimal places.

a)

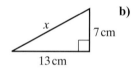

b)

c)

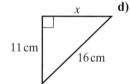

d)

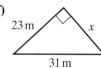

e)

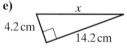

f)

g)

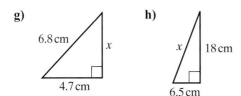

6.8 cm x 4.7 cm

h)

x 18 cm 6.5 cm

2 Ann can walk home from school along two roads or along a path across a field.

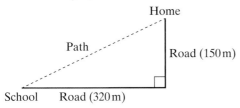

Home

Path

Road (150 m)

School   Road (320 m)

How much shorter is her journey if she takes the path across the field?

3 This network is made of wire.

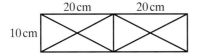

20 cm   20 cm

10 cm

What is the total length of wire?

4 A tangent is drawn from a point P to meet the circle, centre O, at the point T such that TP = 12.8 cm and PT̂O is a right angle. Given that the distance OP = 16.5 cm, calculate the radius of the circle.

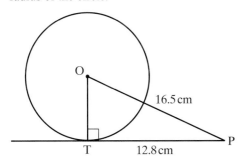

O

16.5 cm

T   12.8 cm   P

5 A prism has a uniform cross-section in the shape of a triangle ABC, right angled at B and in which AC = 5.6 cm and AB = 3.4 cm. The length of the prism is 14.5 cm. Calculate the volume of the prism.

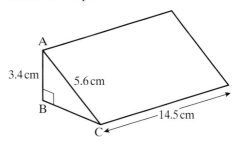

A

3.4 cm   5.6 cm

B

C   14.5 cm

EXERCISE 27.3H

Work out whether or not each of these triangles is right-angled.
Show your working.

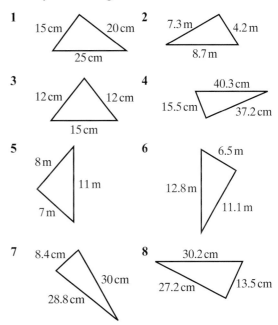

1 15 cm   20 cm   25 cm

2 7.3 m   4.2 m   8.7 m

3 12 cm   12 cm   15 cm

4 40.3 cm   15.5 cm   37.2 cm

5 8 m   11 m   7 m

6 6.5 m   12.8 m   11.1 m

7 8.4 cm   30 cm   28.8 cm

8 30.2 cm   27.2 cm   13.5 cm

EXERCISE 27.4H

1 Find the coordinates of the midpoint of each of the lines in the diagram.

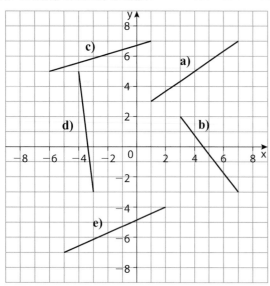

2 Find the coordinates of the midpoint of the line joining each of these pairs of points. Try to do them without plotting the points.
   a) A(3, 7) and B(−5, 7)
   b) C(2, 1) and D(8, 5)
   c) E(3, 7) and F(8, 2)
   d) G(1, 6) and H(9, 3)
   e) I(−7, 1) and J(3, 6)
   f) K(−5, −6) and L(−7, −3)

EXERCISE 27.5H

1 OABCDEFG is a cuboid.
   F is the point (5, 7, 3).

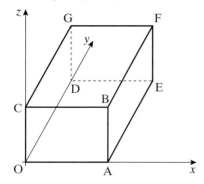

Write down the coordinates of
   a) point A.   **b)** point B.
   c) point C.   **d)** point D.
   e) point E.   **f)** point G.

2 OABCV is a pyramid with a rectangular base. V is directly above the centre of the base, N. OA = 8 units, AB = 10 units and VN = 7 units.

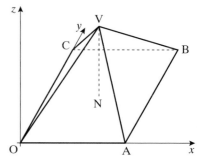

Write down the coordinates of
   a) point A.   **b)** point B.
   c) point C.   **d)** point N.
   e) point V.

3 OABCDEFG is a cuboid.
   M is the midpoint of BF and N is the midpoint of GF.
   OA = 6 units, OC = 5 units and OD = 3 units.

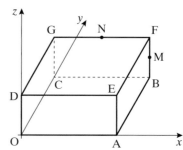

   a) Write down the coordinates of
      (i) point B.
      (ii) point F.
      (iii) point G.
      (iv) point M.
      (v) point N.
   b) **(i)** The point $(6, 2\frac{1}{2}, 0)$ is the midpoint of which edge?
      (ii) The point $(0, 2\frac{1}{2}, 1\frac{1}{2})$ is the centre of which face?

28 → TRANSFORMATIONS

EXERCISE 28.1H

1 Draw a pair of axes and label them
−2 to 4 for *x* and *y*.
   a) Draw a triangle with vertices at
      (1, 1), (1, 3) and (0, 3). Label it A.
   b) Reflect triangle A in the line *x* = 2.
      Label it B.
   c) Reflect triangle A in the line *y* = *x*.
      Label it C.
   d) Reflect triangle A in the line *y* = 2.
      Label it D.

2 Draw a pair of axes and label them
−3 to 3 for *x* and *y*.
   a) Draw a triangle with vertices at (−1, 1),
      (−1, 3) and (−2, 3). Label it A.
   b) Reflect triangle A in the line $x = \frac{1}{2}$.
      Label it B.
   c) Reflect triangle A in the line *y* = *x*.
      Label it C.
   d) Reflect triangle A in the line
      y = −*x*. Label it D.

3 For each part
   • copy the diagram.
   • reflect the shape in the mirror line.
   a) 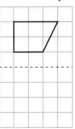 **b)**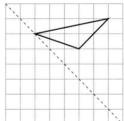

   c)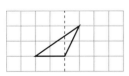

4 Describe fully the single transformation that
maps
   a) shape A on to shape B.
   b) shape A on to shape C.
   c) shape B on to shape D.

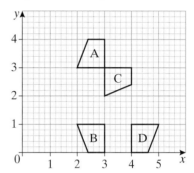

5 Describe fully the single transformation that
maps
   a) triangle A on to triangle B.
   b) triangle A on to triangle C.
   c) triangle E on to triangle F.

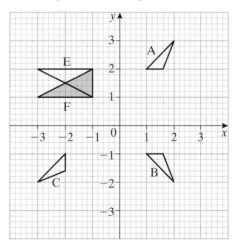

EXERCISE 28.2H

1 Copy the diagram.
   a) Rotate trapezium A through 180° about the origin. Label it B.
   b) Rotate trapezium A through 90° clockwise about the point (0, 1). Label it C.
   c) Rotate trapezium A through 90° anticlockwise about the point (−1, 1). Label it D.

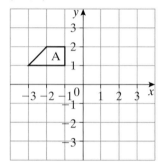

2 Copy the diagram.
   a) Rotate flag A through 90° clockwise about the origin. Label it B.
   b) Rotate flag A through 90° anticlockwise about the point (1, −1). Label it C.
   c) Rotate flag A through 180° about the point (0, −1) Label it D.

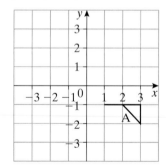

3 Draw a pair of axes and label them −4 to 4 for x and y.
   a) Draw a triangle with vertices (1, 1), (2, 1) and (2, 3). Label it A.
   b) Rotate triangle A through 90° anticlockwise about the origin. Label it B.
   c) Rotate triangle A through 180° about the point (2, 1). Label it C.
   d) Rotate triangle A through 90° clockwise about the point (−2, 1). Label it D.

4 Copy the diagram. Rotate the triangle through 180° about the point C.

5 Copy the diagram. Rotate the shape through 90° clockwise about the point O.

6 Copy the diagram. Rotate the flag through 150° clockwise about the point A.

7 Describe fully the single transformation that maps
   a) trapezium A on to trapezium B.
   b) trapezium A on to trapezium C.
   c) trapezium A on to trapezium D.

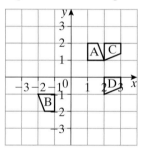

8 Describe fully the transformation that maps
   a) flag A on to flag B.
   b) flag A on to flag C.
   c) flag A on to flag D.

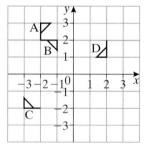

9 Describe fully the single transformation that maps triangle A on to triangle B.

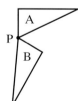

10 Describe fully the single transformation that maps
   a) triangle A on to triangle B.
   b) triangle A on to triangle C.
   c) triangle A on to triangle D.
   d) triangle A on to triangle E.
   e) triangle A on to triangle F.
   Hint: Some of these transformations are reflections.

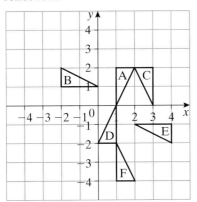

EXERCISE 28.3H

1 Draw a pair of axes and label them -2 to 6 for x and y.
   a) Draw a triangle with vertices at (1, 1), (1, 2), and (4, 1). Label it A.
   b) Translate A by vector $\begin{pmatrix} 1 \\ 3 \end{pmatrix}$. Label it B.
   c) Translate A by vector $\begin{pmatrix} -3 \\ 4 \end{pmatrix}$. Label it C.
   d) Translate A by vector $\begin{pmatrix} -2 \\ -3 \end{pmatrix}$. Label it D.

2 Draw a pair of axes and label them -3 to 5 for x and y.
   a) Draw a triangle with vertices at (2, 1), (2, 3) and (3, 1). Label it A.
   b) Translate A by vector $\begin{pmatrix} 2 \\ 1 \end{pmatrix}$. Label it B.
   c) Translate A by vector $\begin{pmatrix} -5 \\ -3 \end{pmatrix}$. Label it C.
   d) Translate A by vector $\begin{pmatrix} 2 \\ -4 \end{pmatrix}$. Label it D.

3 Describe the single transformation that maps
   a) triangle A on to triangle B.
   b) triangle A on to triangle C.
   c) triangle A on to triangle D.
   d) triangle B on to triangle D.

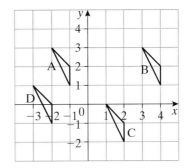

4 Describe the single transformation that maps
   a) shape A on to shape B.
   b) shape A on to shape C.
   c) shape A on to shape D.
   d) shape D on to shape E.
   e) shape A on to shape F.
   f) shape E on to shape G.
   g) shape B on to shape H.
   h) shape H on to shape F.
   Hint: Not all the transformations are translations.

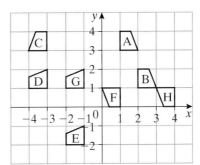

EXERCISE 28.4H

1 Draw a pair of axes and label them 0 to 6 for both x and y.
   a) Draw a triangle with vertices at (0, 6), (3, 6) and (3, 3). Label it A.
   b) Enlarge triangle A by scale factor $\frac{1}{3}$, with the origin as the centre of enlargement. Label it B.
   c) Describe fully the single transformation that maps triangle B on to triangle A.

2 Draw a pair of axes and label them 0 to 6 for both x and y.

   a) Draw a triangle with vertices at (5, 2), (5, 6) and (3, 6). Label it A.

   b) Enlarge triangle A by scale factor $\frac{1}{2}$, with centre of enlargement (3, 2). Label it B.

   c) Describe fully the single transformation that maps triangle B on to triangle A.

3 Draw a pair of axes and label them 0 to 8 for both x and y.

   a) Draw a triangle with vertices at (2, 1), (2, 3), (3, 2). Label it A.

   b) Enlarge triangle A by scale factor $2\frac{1}{2}$, with the origin as the centre of enlargement. Label it B.

   c) Describe fully the single transformation that maps triangle B on to triangle A.

4 Draw a pair of axes and label them 0 to 7 for both x and y.

   a) Draw a trapezium with vertices at (1, 2), (1, 3), (2, 3) and (3, 2). Label it A.

   b) Enlarge trapezium A by scale factor 3, with centre of enlargement (1, 2). Label it B.

   c) Describe fully the single transformation that maps trapezium B on to trapezium A.

5 Describe fully the single transformation that maps

   a) triangle A on to triangle B.

   b) triangle B on to triangle A.

   c) triangle A on to triangle C.

   d) triangle C on to triangle A.

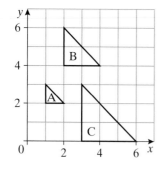

6 Describe fully the single transformation that maps

   a) flag A on to flag B.

   b) flag B on to flag C.

   c) flag B on to flag D.

   d) flag B on to flag E.

   e) flag F on to flag G.

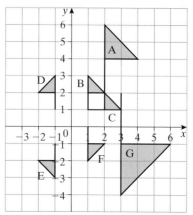

Hint: Not all the transformations are enlargements.

7 Draw a pair of axes and label them −4 to 4 for x and y.

   a) Draw a triangle with vertices at (2, 1), (2, 2) and (4, 2). Label it A.

   b) Reflect A in the line $y = 0$. Label it B.

   c) Reflect A in the line $x = 1$. Label it C.

   d) Rotate B by 90° about the origin. Label it D.

   e) Enlarge A by scale factor $\frac{1}{2}$, with the origin as the centre of enlargement. Label it E.

8 Copy the diagram.

   a) Rotate shape A through 90° clockwise about the origin. Label it B.

   b) Rotate shape A through 180° about (2, 2). Label it C.

   c) Enlarge shape A by scale factor $\frac{1}{2}$, with centre of enlargement (−2, 0). Label it D.

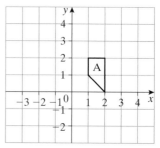

29 → MEASURES

EXERCISE 29.1H

1 Change these measures to the units shown.
   a) 25 cm to mm
   b) 24 m to cm
   c) 1.36 cm to mm
   d) 15.1 cm to mm
   e) 0.235 m to mm

2 Change these measures to the units shown.
   a) 2 m^2 to cm^2
   b) 3 cm^2 to mm^2
   c) 1.12 m^2 to cm^2
   d) 0.05 cm^2 to mm^2
   e) 2 m^2 to mm^2

3 Change these measures to the units shown.
   a) 8000 mm^2 to cm^2
   b) 84 000 mm^2 to cm^2
   c) 2 000 000 cm^2 to m^2
   d) 18 000 000 cm^2 to m^2
   e) 64 000 cm^2 to m^2

4 Change these measures to the units shown.
   a) 32 cm^3 to mm^3
   b) 24 m^3 to cm^3
   c) 5.2 cm^3 to mm^3
   d) 0.42 m^3 to cm^3
   e) 0.02 cm^3 to mm^3

5 Change these measures to the units shown.
   a) 5 200 000 cm^3 to m^3
   b) 270 000 mm^3 to cm^3
   c) 210 cm^3 to m^3
   d) 8.4 m^3 to mm^3
   e) 170 mm^3 to cm^3

6 Change these measures to the units shown.
   a) 36 litres to cm^3
   b) 6300 ml to litres
   c) 1.4 litres to ml
   d) 61 ml to litres
   e) 5400 cm^3 to litres

EXERCISE 29.2H

1 Copy and complete each of these statements.
   a) A length given as 4.3 cm, to 1 decimal place, is between .... cm and .... cm.
   b) A capacity given as 463 ml, to the nearest millilitre, is between .... ml and .... ml.
   c) A time given as 10.5 seconds, to the nearest tenth of a second, is between .... seconds and .... seconds.
   d) A mass given as 78 kg, to the nearest kilogram, is between .... kg and .... kg.
   e) An area given as 5.5 m^2, to 1 decimal place, is between .... m^2 and .... m^2.

2 The number of people attending a football match was given as 24 000, to the nearest thousand.
   What was the least number of people that could have been at the match?

3 Kerry measures her height as 142 cm, to the nearest centimetre.
   Write down the two values between which her height must lie.

4 The height of a desk is stated as 75.0 cm, to 1 decimal place.
   Write down the two values between which its height must lie.

5 Rashid measures the thickness of a piece of plywood as 7.83 mm, to 2 decimal places.
   Write down the two values between which its thickness must lie.

6 The sides of this triangle are given in centimetres, correct to 1 decimal place.

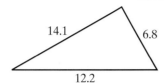

a) Write down the shortest and longest possible length of each side.
b) Write down the shortest and longest possible length of the perimeter.

7 John has two pieces of string.
He measures them as 125 mm and 182 mm, to the nearest millimetre.
He puts the two pieces end to end.
What is the shortest and longest that their combined lengths can be?

8 Mel and Mary both buy some apples. Mel buys 3.5 kg and Mary buys 4.2 kg.
Both weights are correct to the nearest tenth of a kilogram.
a) What is the smallest possible difference between the amounts they have bought?
b) What is the largest possible difference between the amounts they have bought?

EXERCISE 29.3H

1 Rewrite each of these statements using sensible values for the measurements.
a) My mass is 78.32 kg.
b) It takes Katriona 16 minutes and 15.6 seconds to walk to school.
c) The distance to London from Sheffield is 161.64 miles.
d) The length of our classroom is 5 metres 14 cm 3 mm.
e) My water jug hold 3.02 litres.

2 Give your answer to each of these questions to a sensible degree of accuracy.
a) Estimate the length of this line.

_____

b) Estimate the size of this angle.

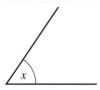

c) A rectangle is 2.3 cm long and 4.5 cm wide. Find the area of the rectangle.
d) The diameter of a circle is 8 cm. Calculate its circumference.
e) The volume of a cube is 7 cm³. Find the length of an edge.
f) An angle in a pie chart is found by working out $\frac{4}{7} \times 360°$. Find the angle.
g) A coach travels 73 miles at an average speed of 33 mph. How long does it take?
h) Six friends share £14 between them. How much does each one get?

EXERCISE 29.4H

1 A boat travels a distance of 24 km in a time of
   3 hours.
   Calculate its average speed.

2 A car covers 197 miles on a motorway in
   3 hours.
   Calculate the average speed. Give your answer
   to 1 decimal place.

3 Anne walks at an average speed of 3.5 km/h
   for 2 hours 30 minutes.
   How far does she walk?

4 How long will it take a boat sailing at 12 km/h
   to travel 64 km?

5 The density of a rock is 9.3 g/cm^3. Its volume
   is 60 cm^3.
   What is its mass?

6 Calculate the density of a piece of metal with
   mass 300 g and volume 84 cm^3.
   Give your answer to a suitable degree of
   accuracy.

7 A man walks 10 km in 2 hours 15 minutes.
   What is his average speed in km/h?
   Give your answer to a suitable degree of
   accuracy.

8 Calculate the mass of a stone of volume
   46 cm^3 and density 7.6 g/cm^3.

9 Copper has a density of 8.9 g/cm^3. Calculate the
   volume of a block of copper of mass 38 g.
   Give your answer to a suitable degree of
   accuracy.

10 What is the density of gas if a mass of 32 kg
    occupies a volume of 25 m^3?
    Give your answer to a suitable degree of
    accuracy.

11 A small town in America has a population of
    235 and covers an area of 35 km^2.
    Find the population density (number of people
    per square kilometre) of the town.

12 A coach left the coach station at 0910 and
    travelled 72 km in 90 minutes, arriving at
    Carter castle.

    a) Calculate its average speed.

    The coach stopped at the castle for $3\frac{1}{2}$ hours
    and then travelled back at an average speed of
    55 km/h.

    b) What time did it arrive back at the coach
        station?
        Give your answer to the nearest minute.

EXERCISE 30.1H

1 Two points, A and B, are 7 cm apart.
Draw the locus of points that are equidistant from A and B.

2 A badger will never go further than 3 miles from its home.
Draw a scale diagram to show the regions where the badger might go looking for food.

3 Draw an equilateral triangle, ABC, of side 6 cm.
Shade the region of points inside the triangle which are nearer to AB than to AC.

4 A rectangular garden measures 8 m by 6 m.
A fence is built from F, at a right angle across the garden.
Draw a scale diagram and construct the line of the fence.

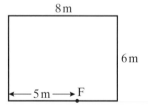

5 Draw a square, ABCD, of side 5 cm.
Draw the locus of points inside the square which are more than 3 cm from A.

6 Zeke is walking across a field.
He notices a bull starting to chase after him.
He runs the shortest distance to the hedge.
Copy the diagram and draw the path that Zeke should run.

• Zeke

7 Draw a rectangle, PQRS, with sides PQ = 7 cm and QR = 5 cm.
Shade the region of points that are closer to P than to Q.

8 Draw an angle of 80°.
Construct the bisector of the angle.

9 An office is a rectangle measuring 16 m by 12 m. There are two electricity points in the office at opposite corners of the room. The vacuum cleaner has a wire 10 m long.
Make a scale drawing to show how much of the room can be cleaned.

10 Make another scale drawing of the office in question **9**.
Shade the locus of points which are equidistant from the two electricity points.
Use this locus to work out the length of wire needed for the vacuum cleaner to reach everywhere in the office.

EXERCISE 30.2H

1 Draw a point and label it P.
Draw the region of points that are less than
4 cm from P and more than 6 cm from P.

2 A rectangular garden measures 20 m by 12 m.
A tree is to be planted so that it is more than 4 m
from each corner of the garden.
Make a scale drawing to find the area where
the tree can be planted.

3 Two points, A and B, are 5 cm apart.
Find the region that is less than 3 cm from A
and more than 4 cm from B.

4 Make an accurate drawing of a triangle, PQR,
where PQ = 6 cm, P = 40° and Q = 35°.
Find the point X, which is 2 cm from R and
equidistant from P and Q.

5 Two coastguard stations, A and B, are 20 km
apart on a straight coastline.
The coastguard at A knows that a ship is
within 15 km of him.
The coastguard at B knows that the same ship
is within 10 km of him.
Make a scale drawing to show the region
where the ship could be.

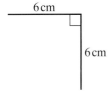

6 Two lines, each 6 cm long, join to form a right
angle.
Draw the region of points which are less than
3 cm from these lines.

7 Two points, P and Q, are 7 cm apart.
Find the points which are the same distance
from P and Q and are also within 5 cm of Q.

8 The diagram shows three coastguard stations,
C, D and E.
A ship is within 25 km of C and closer to DE
than DC.
Find the region where the ship could be.

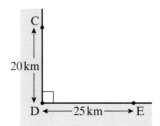

9 A garden is a rectangle, ABCD, with
AB = 5 m and BC = 3 m.
A new flower bed is to be made in the garden.
It must be more than 2 m from A and less than
1.5 m from CD.
Make a scale drawing to show where the
flower bed could be.

10 EFG is a triangle with EF = 6 cm, FG = 8 cm
and EG = 10 cm.
Draw the perpendicular from F to EG.
Indicate the points on this line that are more
than 7 cm from G.

31 ➡ AREAS, VOLUMES AND 2-D REPRESENTATION

EXERCISE 31.1H

Find the circumferences of circles with these diameters.

1 8 cm	**2** 17 cm	**3** 39.2 cm
4 116 mm	**5** 5.1 m	**6** 6.32 m
7 14 cm	**8** 23 cm	**9** 78 mm
10 39 mm	**11** 4.4 m	**12** 2.75 m

EXERCISE 31.2H

1 Find the areas of circles with the following radii.

a)	17 cm	**b)**	23 cm
c)	67 cm	**d)**	43 mm
e)	74 mm	**f)**	32 cm
g)	58 cm	**h)**	4.3 cm
i)	8.7 cm	**j)**	47 m
k)	1.9 m	**l)**	2.58 m

2 Find the areas of circles with the following diameters.

a)	18 cm	**b)**	28 cm
c)	68 cm	**d)**	38 mm
e)	78 mm	**f)**	58 cm
g)	46 cm	**h)**	6.4 cm
i)	7.6 cm	**j)**	32 m
k)	3.4 m	**l)**	4.32 m

EXERCISE 31.3H

Find the area of each of these shapes.
Break them down into rectangles and right-angled triangles first.

1

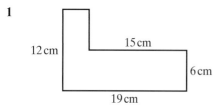

2

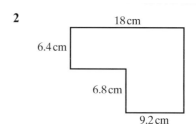

3

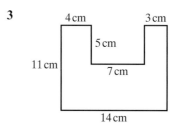

4

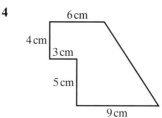

5

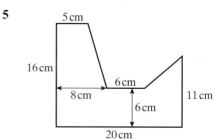

6

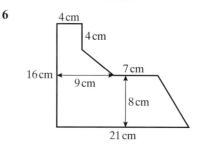

EXERCISE 31.4H

Find the volume of each of these shapes.

1

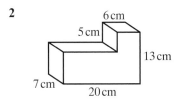

8 cm
5 cm
7 cm
6 cm
14 cm

2

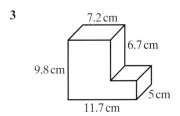

6 cm
5 cm
13 cm
7 cm
20 cm

3

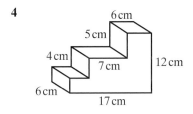

7.2 cm
6.7 cm
9.8 cm
5 cm
11.7 cm

4

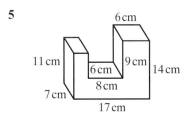

6 cm
5 cm
4 cm
7 cm
12 cm
6 cm
17 cm

5

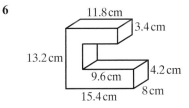

6 cm
11 cm
6 cm
9 cm
14 cm
8 cm
7 cm
17 cm

6

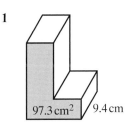

11.8 cm
3.4 cm
13.2 cm
4.2 cm
9.6 cm
8 cm
15.4 cm

EXERCISE 31.5H

Find the volume of each of these prisms.

1

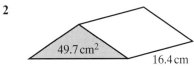

97.3 cm²
9.4 cm

2

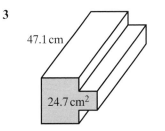

49.7 cm²
16.4 cm

3

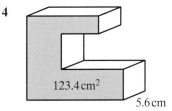

47.1 cm
24.7 cm²

4

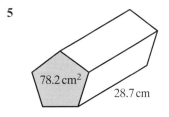

123.4 cm²
5.6 cm

5

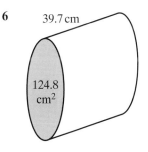

78.2 cm²
28.7 cm

6

39.7 cm
124.8 cm²

EXERCISE 31.6H

Find the volumes of cylinders with these dimensions.

1 Radius 7 cm and height 29 cm

2 Radius 13 cm and height 27 cm

3 Radius 25 cm and height 80 cm

4 Radius 14 mm and height 35 mm

5 Radius 28 mm and height 8 mm

6 Radius 0.6 mm and height 5.1 mm

7 Radius 1.7 m and height 5 m

8 Radius 2.6 m and height 3.4 m

EXERCISE 31.7H

1 Find the curved surface areas of cylinders with these dimensions.
  a) Radius 9 cm and height 16 cm
  b) Radius 13 cm and height 21 cm
  c) Radius 27 cm and height 12 cm
  d) Radius 17 mm and height 35 mm
  e) Radius 12 mm and height 6 mm
  f) Radius 3.7 mm and height 63 mm
  g) Radius 1.9 m and height 19 m
  h) Radius 2.7 m and height 4.3 m

2 Find the total surface areas of cylinders with these dimensions.
  a) Radius 8 cm and height 11 cm
  b) Radius 17 cm and height 28 cm
  c) Radius 29 cm and height 15 cm
  d) Radius 32 mm and height 8 mm
  e) Radius 35 mm and height 12 mm
  f) Radius 3.9 mm and height 45 mm
  g) Radius 0.8 m and height 7 m
  h) Radius 2.9 m and height 1.7 m

EXERCISE 31.8H

Draw the plan view, front elevation and side elevation of each of these objects.

1

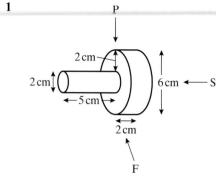

2

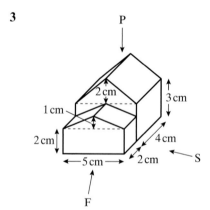

3

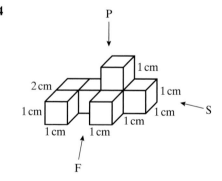

4

EXERCISE 32.1H

1 Find the lengths marked *a*, *b*, *c*, *d*, *e* ,*f*, *g* and *h* in these diagrams.

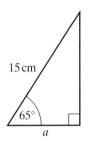
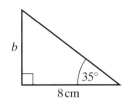
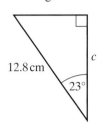
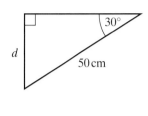

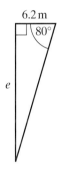
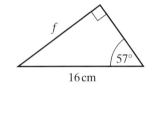
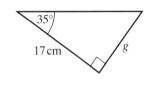
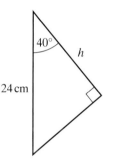

2 The diagram shows the side view of a waste bin.
 a) Find the width, *w* cm, of the bin.
 b) Find the height, *h* cm, of the bin.

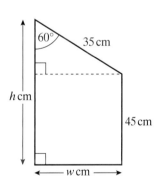

3 a) Find the height, *a* cm, of the triangle.
 b) Find the length, *b* cm.
 c) Find the length, *c* cm.
 d) Use your answers to parts **a)**, **b)** and **c)** to find the
   area of the triangle.

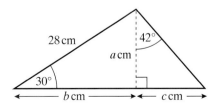

1   Find the lengths marked a, b, c, d, e, f, g and h in these diagrams.

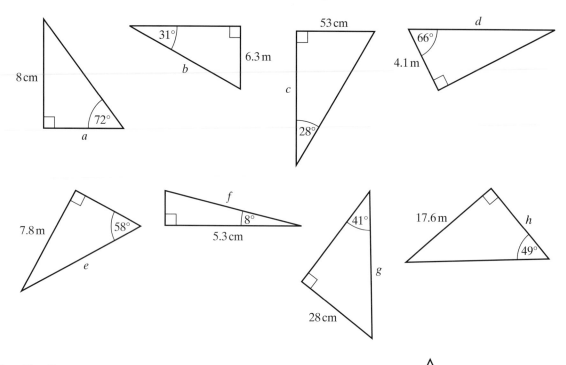

2   The diagram represents a step ladder.
    The two sections of the ladder are opened to 30°.
    The feet of the two parts are 1.2 m apart.
    Calculate the length, l, of each section of the ladder.

3   **a)**   Find the base, a cm, of the triangle.
    b)   Use the base you found in part **a)** to find the area
            of the triangle.

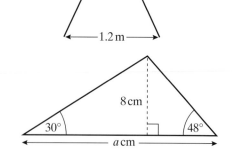

EXERCISE 32.3H

1 Find the angles marked a, b, c, d, e, f, g and h in these diagrams.

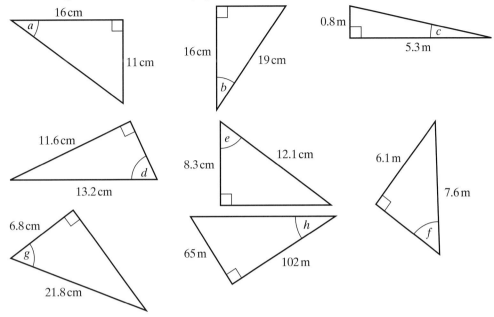

2 In the diagram, ABC is a straight line and BDE is a
straight line perpendicular to it. It is given that
AD = 36 m, BC = 49 m, $D\hat{A}B$ = 43° and
$E\hat{C}B$ = 54°.
Calculate the length of DE.

3 A ramp is to be made to improve the access to a
building. The height of the step into the building
is 18 cm and there is room for a ramp that is 85 cm
along the ground. Find the angle, denoted by x,
that the ramp makes with the ground.

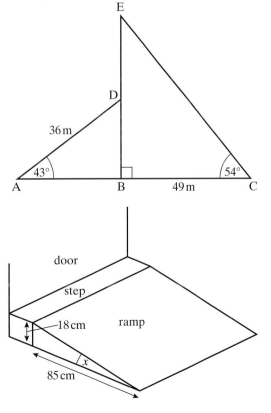

EXERCISE 33.1H

1 Find the area of each of these triangles.

a)

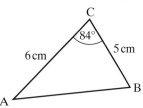

b)

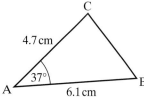

c)

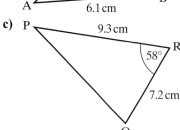

d)

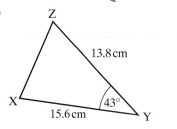

2 a) In triangle ABC, $b = 9$ cm, $c = 16$ cm and the area is 60.4 cm². Find the size of $\hat{CAB}$.

b) In triangle PQR, $q = 6.4$ cm, $r = 7.8$ cm and the area is 12.1 cm². Find the size of $\hat{RPQ}$.

c) In triangle XYZ, $x = 23.7$ cm, $y = 16.3$ cm and the area is 184 cm². Find the size of $\hat{YZX}$.

3 a) In triangle ABC, $b = 20$ cm, angle C is 27° and the area is 145.3 cm². Find the length of side BC.

b) In triangle XYZ, $x = 9.3$ cm, angle Z is 94° and the area is 34.3 cm². Find the length of side XZ.

EXERCISE 33.2H

1 Find the size of each of the sides and angles not given in these diagrams.

a)

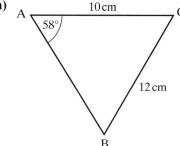

b)

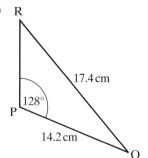

c)

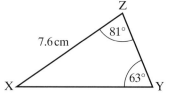

d)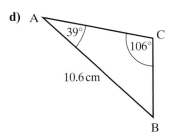

2 In triangle PQR, $P\hat{Q}R$ is 38°, side PR is 8.3 cm and side PQ is 12 cm.
Calculate the size of the largest angle of the triangle.

EXERCISE 33.3H

1 Find the size of each of the sides and angles marked in these diagrams.

a)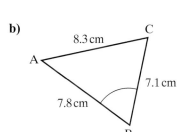

b)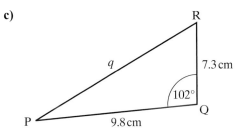

c)

d)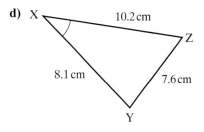

2 Find the size of each of the angles in these diagrams.

a)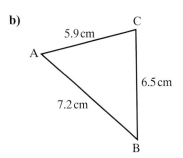

b)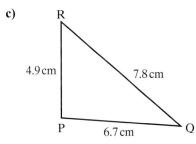

c)

d)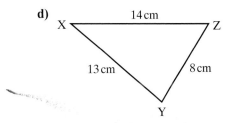

3 ABCDEFGH is a cuboid.
BDE is a triangle contained within the cuboid.

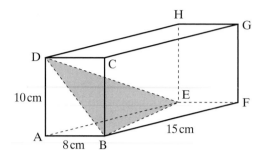

Calculate the size of these angles.
a) Angle BDE
b) Angle BED

EXERCISE 33.4H

1 **a)** Draw accurately the graph of $y = \sin \theta$
for values of θ from $-180°$ to $180°$.
b) For what values of θ in this range does
$\sin \theta = 0.7$?

2 **a)** Draw accurately the graph of $y = \cos \theta$
for values of θ from $0°$ to $360°$.
b) For what values of θ in this range does
$\cos \theta = -0.6$?

3 **a)** Sketch the graph of $y = \sin \theta$ for values
of θ from $0°$ to $360°$.
b) Use your sketch and your calculator to
find all the values of θ for which
$\sin \theta = -0.8$.

4 **a)** Sketch the graph of $y = \cos \theta$ for values
of θ from $-180°$ to $360°$.
b) Use your sketch and your calculator to
find all the solutions of the equation
$\cos \theta = -0.3$.

5 One solution to $\sin \theta = 0.6$ is approximately $37°$.
Using only the symmetry of the sine curve,
find the other angles between $-180°$ and $540°$
that also satisfy the equation $\sin \theta = 0.6$.

6 **a)** Draw accurately the graph of $y = \tan \theta$
for values of θ from $-180°$ to $360°$.
b) Find, from your graph, the angles for
which $\tan \theta = -1.5$.

EXERCISE 33.5H

1 Find the amplitude and the period of each of
these curves.
a) $y = 2 \sin \theta$ **b)** $y = \cos 4\theta$
c) $y = 5 \sin 2\theta$ **d)** $y = 3 \sin 0.8\theta$
e) $y = 4 \cos 5\theta$ **f)** $y = 7 \cos 0.6\theta$

2 Draw the graph of $y = 2 \cos \theta$ for values of θ
from $0°$ to $360°$.

3 Draw the graph of $y = \sin 3\theta$ for values of θ
from $-180°$ to $180°$.

4 Sketch the graph of $y = \cos 5\theta$ for values of θ
from $0°$ to $360°$.

5 Sketch the graph of $y = 1.5 \sin \theta$ for values of
θ from $-180°$ to $180°$.

6 Find the solutions of $\cos 4\theta = -0.6$ between
$0°$ and $360°$.

EXERCISE 34.1H

Throughout this exercise, letters in algebraic expressions represent lengths and numbers have no dimensions.

1 Which of these expressions could be a length?
   a) πd
   b) $6x^2$
   c) $r(\pi + 4)$

2 Which of these expressions could be an area?
   a) $3ab(4c + d)$
   b) $20bh$
   c) $a^2 + 6b^2$

3 Which of these expressions could be a volume?
   a) $2\pi r^3$
   b) $4b(c + d)$
   c) $\pi r^2 h$

4 State whether each of these expressions represents a length, an area, a volume or none of these.
   a) πrh
   b) $2\pi r(r^2 + 2h)$
   c) $\frac{1}{2}(a + b)hp$
   d) $\dfrac{12x^3}{ab}$

5 Each of the following quantities has a particular number of dimensions. Give the dimensions of **each** quantity.
   a) The volume of a bucket
   b) The radius of a circle
   c) The cross-sectional area of a cylinder
   d) The length of a pencil
   e) The distance to the moon
   f) The area of a football pitch

EXERCISE 34.2H

1 Find the arc length of each of these sectors.
   Give your answers to the nearest millimetre.

   a)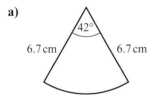
   b)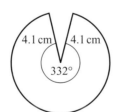
   c)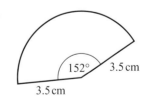

2 Find the area of each of the sectors in question **1**.

3 Find the sector angle of each of these sectors.
Give your answers to the nearest degree.

a)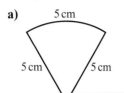
5 cm
5 cm 5 cm

b)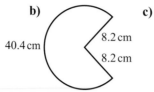
40.4 cm
8.2 cm
8.2 cm

c)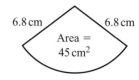
6.8 cm 6.8 cm
Area =
45 cm²

d)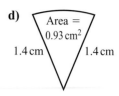
Area =
0.93 cm²
1.4 cm 1.4 cm

4 Find the radius of each of these sectors.

a)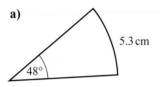
5.3 cm
48°

b)
8.5 cm
167°

c)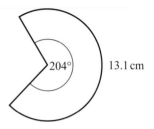
204° 13.1 cm

5 A sector of a circle of radius 5.1 cm has an area of 38 cm².
Calculate the angle of the sector and hence find the arc length of the sector.

EXERCISE 34.3H

1 Find the curved surface area of each of these cones.
Give your answers to 3 significant figures.

a)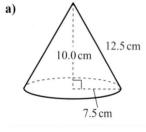
12.5 cm
10.0 cm
7.5 cm

b)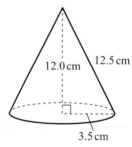
12.5 cm
12.0 cm
3.5 cm

c)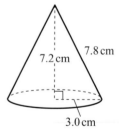
7.8 cm
7.2 cm
3.0 cm

2 Calculate the volume of each of the cones in question **1**.

3 Calculate the volume of each of these pyramids.

a)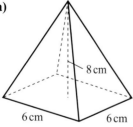
8 cm
6 cm 6 cm

b)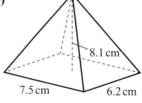
8.1 cm
7.5 cm 6.2 cm

c)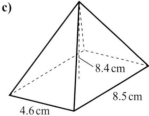
8.4 cm
4.6 cm 8.5 cm

4 A solid cone has a base of radius 6.1 cm and a slant height of 8.4 cm. Calculate its total surface area.

5 A pyramid has a rectangular base of sides 4.5 cm and 6.0 cm. Its volume is 73.8 cm³. Find its height.

6 Find the base radius of cones with these measurements.
  a) Volume 256 cm³, height 7.8 cm
  b) Volume 343 cm³, height 6.5 cm
  c) Volume 192 cm³, height 10.4 cm

7 A sector of a circle is joined to form a cone.
Find the radius of the base of the cone made with each of these sectors.
  a) Radius 8.1 cm, angle 150°
  b) Radius 10.8 cm, angle 315°
  c) Radius 7.2 cm, angle 247°

8 Find the surface area of spheres with these measurements.
  a) Radius 3.6 cm   **b)** Radius 8.5 cm   **c)** Diameter 36 cm

9 Find the volume of each of the spheres in question **8**.

10 Find the radius of spheres with these measurements.
  a) Surface area 650 cm²   **b)** Surface area 270 cm²
  c) Volume 1820 cm³   **d)** Volume 3840 cm³

11 A solid metal sphere of radius 6.5 cm is melted down and recast as a solid cone whose base radius is half its height. Calculate the radius of the cone.

12 A cone has height 3.5 cm and the radius of the base is 1.2 cm. A pyramid has height 2.3 cm and it has the same volume as the cone. Find the area of the base of the pyramid.

13 A sphere has a radius of 3.5 cm. A cone with a volume equal to that of the sphere has a base radius of 4.5 cm. Calculate the height of the cone.

14 A solid metal cone of radius 8.2 cm is melted down and recast as a solid square-based pyramid of the same height as the cone. Calculate the length of the side of the base of the pyramid.

15 The diagram shows a toy made by fixing a cone onto a hemisphere.
The radius of the base of the cone and the radius of the hemisphere are both 4.6 cm.
The overall height of the toy is 9.8 cm.
Find the volume of the toy.

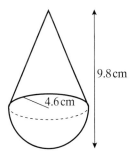

16 Each of the faces of a regular tetrahedron is an equilateral triangle of side $2x$ cm. Show that the surface area of the tetrahedron is $4\sqrt{3}x^2$ cm².

1 Calculate the area of each of the shaded segments.

a)

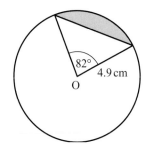

b)

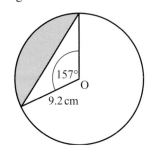

c)

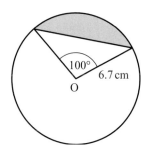

2 A concrete gatepost is in the shape of a cuboid 1.6 m by 30 cm by 30 cm topped with a sphere of radius 12 cm.
Calculate the volume of the gatepost.

3 Calculate the area of each of the shaded major segments.

a)

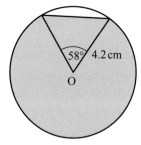

b)

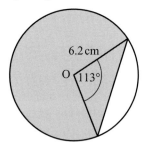

c)

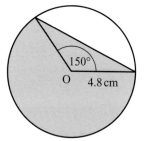

4 A birthday cake has radius 9 cm and height 8 cm. Its top and sides are covered in icing.
Marie is given a slice of the cake. It is a sector of angle 50°.
Calculate the surface area of the icing on Marie's slice of cake.

5 Calculate the perpendicular height of each of these cones and hence find their volumes.

a)

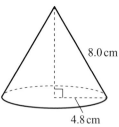

b)

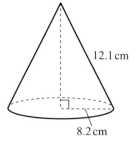

c)

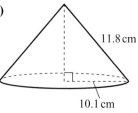

6   A solid cone has a base of radius 7.9 cm and a height of 11.8 cm.
   a)   Calculate its volume.
   b)   Find its slant height and hence its total surface area.

7   The top half (in height) of a cone is removed leaving a frustum.
   a)   Show that the volume of the remaining frustum is $\frac{7}{8}$ of the volume of the original cone.
   b)   What fraction of the original curved surface area of the cone is the curved surface area of the frustum?

8   A podium is a frustum of a cone. The height of the frustum is 1.2 m and the diameters of its top and its base are 0.8 m and 1.4 m respectively.
   a)   Show that the podium is a frustum of a cone of complete height 2.8 m.
   b)   Calculate the volume of the podium.

9   A pyramid has a square base of side 12.8 cm and its volume is 524 cm³.
   a)   Calculate its height.
   b)   Hence show that the sloping edges (which are all of equal length) are 13.2 cm long.

10   A lampshade is made from a piece of parchment.
   a)   Find in the form $\dfrac{k}{\pi}$ the angle of the sector of the inner circle which has been removed.
   b)   Hence calculate the surface area of the lampshade.

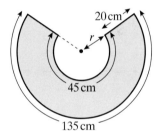

EXERCISE 35.1H

1 Two rectangles are similar.
The ratio of the corresponding sides of the two rectangles is 2 : 5.
The width of the smaller rectangle is 9 cm and the length of the larger one is 45 cm.
Find
a) the length of the smaller rectangle.
b) the width of the larger rectangle.

2 Write down the triangle which is similar to triangle ABC.
Make sure you put the letters in the correct order.

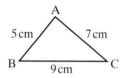
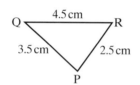

3 For these two triangles, write down the corresponding sides.

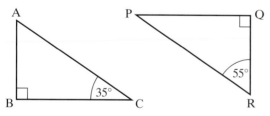

4 Triangle ABC is similar to triangle QRP.

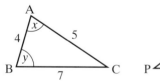
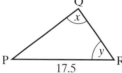

Calculate the lengths of PQ and QR.

5 In triangle ABC, XY is parallel to BC,
XA = 4 cm, YA = 5 cm, BX = 2 cm and
XY = 2 cm.

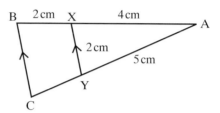

a) Which triangle is similar to triangle ABC?
b) Calculate the lengths of BC and CY.

6 Calculate the lengths x and y in this diagram.

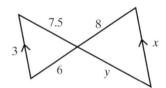

7 Triangles ABC and CBD are similar.
AC = 3 cm, BC = 6 cm and BD = 9 cm.

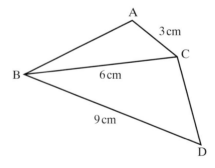

Calculate these lengths.
a) AB
b) CD

8 This is a diagram of a triangle with its top cut off.
AB is parallel to XY.

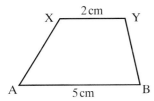

AB = 5 cm and XY = 2 cm.
The height of the piece cut off is 3.5 cm.
Find the height of the complete triangle.

EXERCISE 35.2H

1 Draw a set of axes with the *x*-axis from −8 to 16 and the *y*-axis from −6 to 10.
Plot the points A(−2, −1), B(−2, −4) and C(−7, −1) and join them to form a triangle.
Enlarge the triangle by a scale factor of −2 using the origin as the centre of enlargement.

2 Draw a set of axes with the *x*-axis from −8 to 8 and the *y*-axis from −8 to 8.
Plot the points A(6, 5), B(6, 8) and C(8, 8) and join them to form a triangle.
Enlarge the triangle by a scale factor of −3 using (4, 4) as the centre of enlargement.

3 The diagram shows a quadrilateral, ABCD, and its image A′B′C′D′.

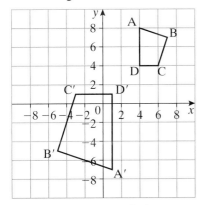

Copy the diagram and find
a) the centre of enlargement.
b) the scale factor.

4 Draw a set of axes with the *x*-axis from −10 to 10 and the *y*-axis from −8 to 8.
Plot the points A(−4, 3), B(−4, −3) and C(−10, −5) and join them to form a triangle.
Enlarge the triangle by a scale factor of $-\frac{1}{2}$ using (2, 3) as the centre of enlargement.

5 The diagram shows a triangle, ABC, and its image A′B′C′.

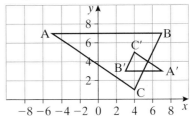

Copy the diagram and find
a) the centre of enlargement.
b) the scale factor.

EXERCISE 35.3H

1 A cube has edges 4 cm long. Another cube has edges 12 cm long.
a) Write down the linear scale factor for the enlargement.
b) Write down the area of
  (i) a face of the small cube.
  (ii) a face of the big cube.
c) **(i)** Write down the area scale factor.
  (ii) What do you notice?
d) Write down the volume of
  (i) the small cube.
  (ii) the big cube.
e) **(i)** Write down the volume scale factor.
  (ii) What do you notice?

2 Triangles ABC and PQR are similar.

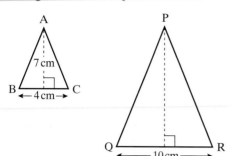

a) What is the linear scale factor of the enlargement?
b) Find the height of triangle PQR.
c) Calculate the area of triangle ABC.
d) Calculate the area of triangle PQR.
e) Write down the ratio of the areas.

3 State the area scale factor and the volume scale factor for each of these linear scale factors.
a) 3
b) 4
c) 8
d) $\frac{1}{2}$
e) $\frac{5}{3}$

4 State the linear scale factor for each of these area scale factors.
a) 49
b) 81
c) 100
d) 2500
e) $\frac{9}{25}$

5 A model boat is built to a scale of 1 : 50. It is 30 cm long.
a) How long is the real boat?
b) The real boat has a mast 8 m high. How high is the mast on the model?

6 A table is 0.6 m wide and the top has an area of 0.56 m². Another table has a similar shaped top. It is 0.9 m wide.
What is the area of the top of this table?

7 A small bottle holds 150 ml of liquid. A similar bottle is twice as tall.
How much liquid does it hold?

8 Shirley has a poster made from a photo she has taken.
The poster is an enlargement of the photo with linear scale factor 8.
The dimensions of the photo are 5 cm by 7 cm.
What is the area of the poster?

9 A vase is 12 cm tall.
Another similar vase is 18 cm tall.
The larger vase has a capacity of 54 cm³.
What is the capacity of the smaller vase?

10 Two cuboids are similar.
The smaller cuboid has edges 4 cm, 5 cm and 8 cm long.
The larger one has a volume of 20 000 cm³.
What is the length of the shortest edge of the larger cuboid?

11 a) Explain clearly why the following triangles are **not** similar.

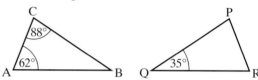

b) Triangles DEF and XYZ are similar. Their corresponding sides are in the ratio 4 : 3. Calculate the length of YZ.

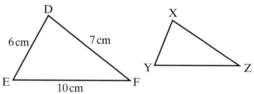

12 a) Explain clearly why the following triangles are similar.

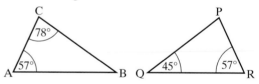

b) Explain clearly why the following triangles are **not** similar.

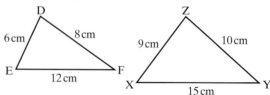

36 → THREE-DIMENSIONAL GEOMETRY

EXERCISE 36.1H

1   Calculate the length of the diagonal of a cuboid measuring 6 cm by 10 cm by 5 cm.

2   The length of the diagonal of a cube is 6.8 cm. Find the length of a side of this cube.

3   In this cuboid, AB = 10 cm, BC = 6 cm and CG = 8 cm.

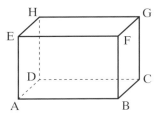

Calculate
a)   angle GDC.
b)   the length of EG.
c)   the length of HB.
d)   angle BHD.

4   A pyramid is 8 cm high and has a square base of side 6 cm.
Its sloping edges are all of equal length.
Calculate the length of a sloping edge.

5   The length of each sloping edge of a square-based pyramid is 12 cm.
The sides of the base are each 10 cm.
Calculate the height of the pyramid.

6   ABCDEF is a triangular wedge.
The faces ABFE, BCDF and ACDE are rectangles.

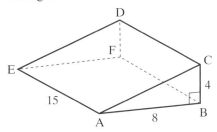

a)   Calculate the length AD.
b)   Calculate DÂC.

7   A is 30 m due south of a church tower.
From A, the elevation of the top of the tower is 51°.
a)   Calculate the height of the tower.
From B, which is due west of the tower, the angle of elevation of the top of the tower is 35°.
b)   Calculate how far B is from the tower.
c)   Calculate the distance AB, assuming that A and B are on the same level as the base of the tower.

8   A pyramid VABCD has a square base ABCD of side 6 cm.
O is the centre of the base.
a)   Show that AO = $\sqrt{18}$ cm.
Angle VAO, the angle between a sloping edge and the base, is 62°.
b)   Calculate the height VO.
c)   Calculate the length of a sloping edge of the pyramid.

9 The pyramid OABCD has a horizontal rectangular base ABCD as shown.
O is vertically above A.

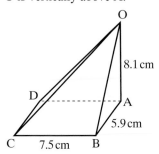

Calculate
a) the length of OB.
b) angle OCB.
c) the length of OC.

⌨ EXERCISE 36.2H

1 A square-based pyramid has sloping edges of length 9.5 cm.
The sloping edges make an angle of 60° with the base.
Calculate the height of the pyramid.

2 A pyramid is 7 cm high and has a square base of side 5 cm.
Its sloping edges are all of equal length.
Calculate the angle between a sloping face and the base.

3 In this cuboid, AB = 12 cm, BC = 5 cm and CG = 6 cm.

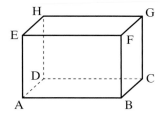

Calculate
a) the length of AC.
b) the length of AG.
c) the angle between AG and the base ABCD.
d) the angle between AG and face BCGF.

4 The length of the diagonal of a cuboid is 9.3 cm.
The height of the cuboid is 5.6 cm.
Calculate the angle between the diagonal and a vertical edge of the cuboid.

5 VABCD is a square-based pyramid of height 8 cm.
Its base ABCD has side 10 cm.
All its sloping edges are equal in length.
M is the midpoint of AB.

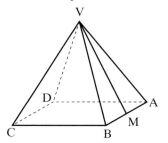

Calculate
a) the angle which VM makes with the base.
b) the length of VM.
c) the length of VA.
d) the angle which VA makes with the base.

6 VABCD is a square-based pyramid of height 8 cm.
Its base ABCD has side 10 cm.
V is directly above A.

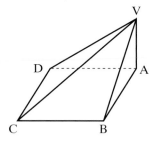

Calculate the lengths of the sloping edges VB, VC and VD, and the angles they make with the base.

7 This 'lean-to' workshop is a prism with a trapezium as its cross-section.

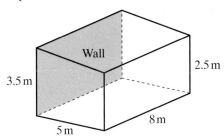

a) Calculate the area of the sloping roof.
b) Calculate the angle between the roof and the wall against which the workshop is built.
c) Calculate the length of the longest piece of wood that will fit in the workshop. (Assume that the door is large enough for it to get in!)

8 The diagonal of a cuboid has length 12.4 cm. It makes an angle of 33° with the base of the cuboid.
a) Calculate the height of the cuboid.
b) The length of the base of the cuboid is 5.8 cm. Calculate its width.

EXERCISE 37.1H

1 Triangles ABC and PQR are congruent.

   a) Write down the size of angle
   - **(i)** BAC
   - **(ii)** BCA
   - **(iii)** PQR
   - **(iv)** RPQ

   b) Which side in triangle PQR is 7.5 cm long?

2 Which triangles are congruent to triangle ABC?

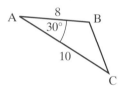
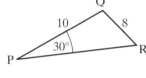
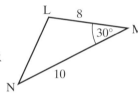
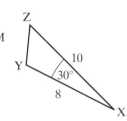

3 'SAS' is shorthand for 'Two sides and the included angle of one triangle are equal to the two sides and the included angle of the other triangle.'
Explain what each of these is shorthand for.
   a) SSS   **b)** ASA   **c)** RHS

4 State whether or not each of these pairs of triangles are congruent.
Give reasons for your answers.

a)
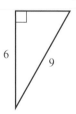

b)
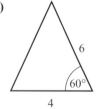

c)
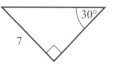

d)
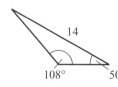

5 In this diagram, AB = BC and $A\hat{B}D = D\hat{B}C$.
Prove that triangle ABD is congruent to triangle CBD.

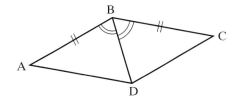

6 In this diagram, AB is parallel to ED and BC = CE.
Prove that triangles ABC and DEC are congruent.

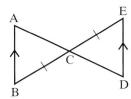

7 a) Sketch a quadrilateral ABCD in which $A\hat{B}C = A\hat{D}C$ and BC is parallel to AD.
Join the vertices A and C to make two triangles.
(You may recognise the shape you have drawn, but do not make any other assumptions about the shape except what you have been told.)
  b) Prove that triangles ABC and CDA are congruent.
  c) What does this prove about the opposite sides of the quadrilateral?

8 a) Sketch an isosceles triangle ABC with AB = AC.
Draw a straight line from the midpoint of BC to A.
  b) Use congruent triangles to prove that this line bisects the angle at A and is perpendicular to the side BC.

9 a) Sketch an equilateral triangle ABC.
The midpoint of AB is X, the midpoint of BC is Y, and the midpoint of AC is Z.
  b) Prove that triangle XYZ is also equilateral.

10 Two triangles have two sides of length 5 cm and 9 cm.
The angle opposite the 5 cm side is 25°.
The two triangles are not congruent.
Sketch the two triangles to show that they are not congruent.

EXERCISE 38.1H

In each of the questions, find the size of the angle or length marked with a lower-case letter.
Give a reason for each step of your work.

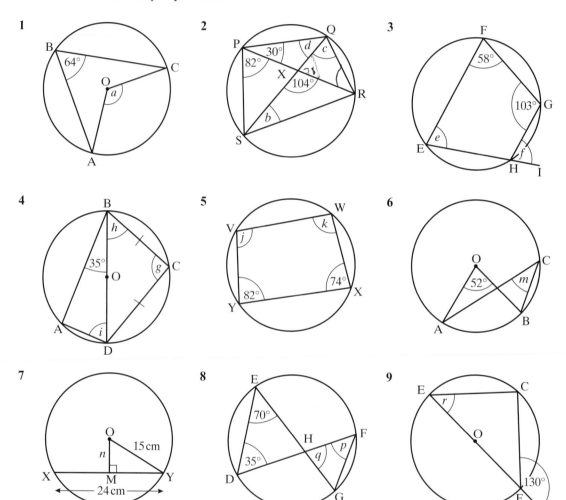

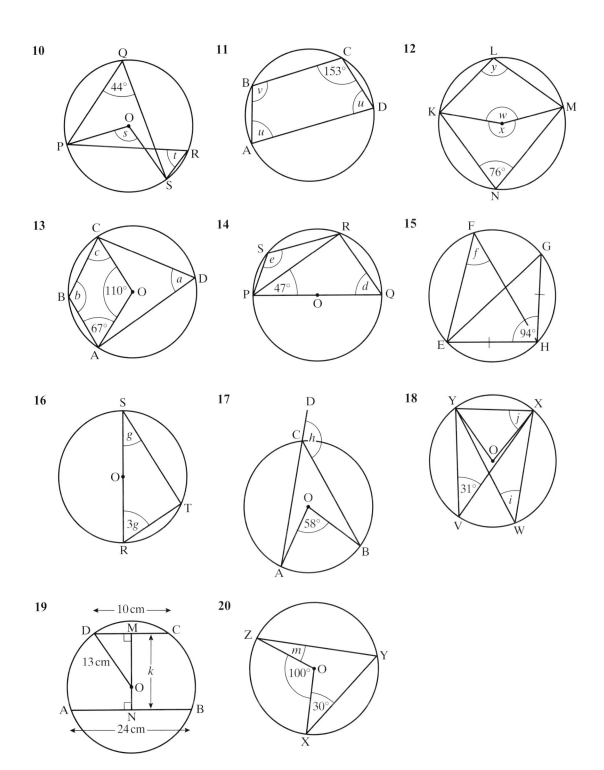

EXERCISE 38.2H

In each of the questions, find the size of the angles marked with a lower-case letter.
Give a reason for each step of your work.

1

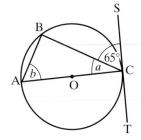

2

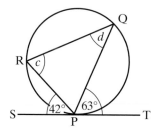

3

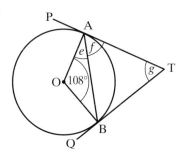

4

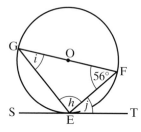

5

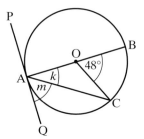

6

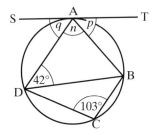

7

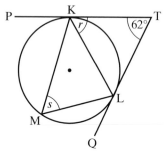

8

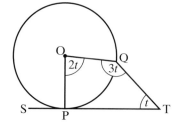

9

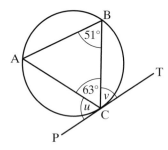

10

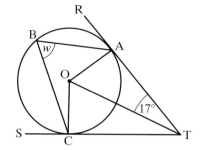

39 → STATISTICAL DIAGRAMS

EXERCISE 39.1H

1 For a project, Rebecca recorded the ages of 100 cars as they passed the school gates one morning. Here are her results.

Age (a years)	$0 \leqslant a < 2$	$2 \leqslant a < 4$	$4 \leqslant a < 6$	$6 \leqslant a < 8$	$8 \leqslant a < 10$	$10 \leqslant a < 12$	$12 \leqslant a < 14$
Frequency	16	23	24	17	12	7	1

    a) Draw a frequency diagram to show these data.
    b) Which of the intervals is the modal class?

2 The manager of a leisure centre recorded the weights of 120 men. Here are the results.

Weight (w kg)	$60 \leqslant w < 65$	$65 \leqslant w < 70$	$70 \leqslant w < 75$	$75 \leqslant w < 80$	$80 \leqslant w < 85$	$85 \leqslant w < 90$
Frequency	4	18	36	50	10	2

    a) Draw a frequency diagram to represent these data.
    b) Which of the intervals is the modal class?
    c) Which of the intervals contains the median value?

3 This frequency diagram shows the times taken by a group of girls to run a race.

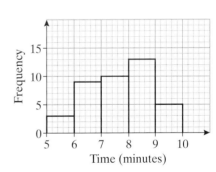

    a) How many girls took longer than 9 minutes?
    b) How many girls took part in the race?
    c) What percentage of the girls took less than 7 minutes?
    d) What is the modal finishing time?
    e) Use the diagram to draw up a grouped frequency table like those in questions **1** and **2**.

EXERCISE 39.2H

1 The table shows the heights of 40 plants.

Height (h cm)	$3 \leqslant h < 4$	$4 \leqslant h < 5$	$5 \leqslant h < 6$	$6 \leqslant h < 7$	$7 \leqslant h < 8$	$8 \leqslant h < 9$
Frequency	1	7	10	12	8	2

Draw a frequency polygon to show these data.

2 The table shows the time taken for a group of children to get from home to school.

Time (t mins)	$0 \leqslant t < 5$	$5 \leqslant t < 10$	$10 \leqslant t < 15$	$15 \leqslant t < 20$	$20 \leqslant t < 25$	$25 \leqslant t < 30$
Frequency	3	15	27	34	19	2

Draw a frequency polygon to show these data.

3 The ages of all of the people under 70 in a small village were recorded in 1989 and 2009.
The results are given in the table below.

Age (a years)	$0 \leqslant a < 10$	$10 \leqslant a < 20$	$20 \leqslant a < 30$	$30 \leqslant a < 40$	$40 \leqslant a < 50$	$50 \leqslant a < 60$	$60 \leqslant a < 70$
Frequency 1989	85	78	70	53	40	28	18
Frequency 2009	50	51	78	76	62	64	56

a) On the same grid, draw a frequency polygon for each year.
b) Use the diagram to compare the distribution of ages in the two years.

EXERCISE 39.3H

1   Bill grows tomatoes. As an experiment, he divided his land into eight plots.
    He used a different amount of fertiliser on each plot.
    The table shows the weight of tomatoes he got from each of the plots.

Amount of fertiliser (g/m²)	10	20	30	40	50	60	70	80
Weight of tomatoes (kg)	36	41	58	60	70	76	75	92

a)   Draw a scatter diagram to show this information.
b)   Describe the correlation shown in the scatter diagram.
c)   The mean of the amount of fertiliser is 45 g/m². Calculate the mean weight of the tomatoes.
d)   Plot the point that has these means as coordinates.
e)   Draw a line of best fit on your scatter diagram.
f)   What weight of tomatoes should Bill expect to get if he used 75 g/m² of fertiliser?

2   The table shows the prices and mileages of seven second-hand cars of the same model.

Price (£)	6000	3500	1000	8500	5500	3500	7000
Mileage	29 000	69 000	92 000	17 000	53 000	82 000	43 000

a)   Draw a scatter diagram to show this information.
b)   Describe the correlation shown in the scatter diagram.
c)   The mean price is £5000. Calculate the mean mileage.
d)   Plot the point that has these means as coordinates.
e)   Draw a line of best fit on your scatter diagram.
f)   Use your line of best fit to estimate
     (i)   the price of this model of car with a mileage of 18 000 miles.
     (ii)   the mileage of this model of car which costs £4000.

3   The heights of 10 daughters, all aged 20, and their fathers are given in the table below.

Height of father (cm)	167	168	169	171	172	172	174	175	176	182
Height of daughter (cm)	164	166	166	168	169	170	170	171	173	177

a)   Draw a scatter diagram to show this information.
b)   Describe the correlation shown in the scatter diagram.
c)   The mean height of the fathers is 172.6 cm. Calculate the mean height of the daughters.
d)   Plot the point that has these means as coordinates.
e)   Draw a line of best fit on your scatter diagram.
f)   Use your line of best fit to estimate the height of a 20-year-old daughter whose father is 180 cm tall.

EXERCISE 40.1H

1 For each of these sets of data
   (i) find the mode. **(ii)** find the median. **(iii)** find the range. **(iv)** calculate the mean.

a)

Score on biased dice	Number of times thrown
1	52
2	46
3	70
4	54
5	36
6	42
Total	300

b)

Number of drawing pins in a box	Number of boxes
98	5
99	14
100	36
101	28
102	17
103	13
104	7
Total	120

c)

Number of snacks per day	Frequency
0	23
1	68
2	39
3	21
4	10
5	3
6	1

d)

Number of letters received on Monday	Frequency
0	19
1	37
2	18
3	24
4	12
5	5
6	2
7	3

e)

x	Frequency
48	23
49	62
50	51
51	58
52	30
53	16

f)

x	Frequency
141	3
142	27
143	66
144	81
145	74
146	35
147	12
148	2

2 Gift tokens cost £1, £5, £10, £20 or £50 each.
The frequency table below shows the numbers of each value of gift token sold in one bookstore on a Saturday.
Calculate the mean value of gift token bought in the bookstore that Saturday.

Price of gift token (£)	1	5	10	20	50
Number of tokens sold	12	34	26	9	1

3 A sample of people were asked how many visits to the cinema they had made in one month.
None of those asked had made more than eight visits to the cinema.
The table below shows the data.
Calculate the mean number of visits to the cinema.

Number of visits	0	1	2	3	4	5	6	7	8
Frequency	136	123	72	41	18	0	5	1	4

EXERCISE 40.2H

1 For each of these sets of data, calculate an estimate of
   (i)   the range.   (ii)   the mean.

a)

Number of trains arriving late each day (x)	Number of days (f)
0–4	19
5–9	9
10–14	3
15–19	0
20–24	1
Total	32

b)

Number of weeds per square metre (x)	Number of square metres (f)
0–14	204
15–29	101
30–44	39
45–59	13
60–74	6
75–89	2

c)

Number of books sold (x)	Frequency (f)
60–64	3
65–69	12
70–74	23
75–79	9
80–84	4
85–89	1

d)

Number of days absent (x)	Frequency (f)
0–3	13
4–7	18
8–11	9
12–15	4
16–19	0
20–23	1
24–27	3

2 The table below gives the number of sentences per chapter in a book.

Number of sentences (x)	$100 \leqslant x < 125$	$125 \leqslant x < 150$	$150 \leqslant x < 175$	$175 \leqslant x < 200$	$200 \leqslant x < 225$
Frequency	1	9	8	5	2

   a)   What is the modal class?
   b)   In which group is the median number of sentences?
   c)   Calculate an estimate of the mean number of sentences.

3 A group of students were asked to estimate the number of beans in a jar.
The results of their estimates are summarised in the table.
Calculate an estimate of the mean number of beans estimated by these students.

Estimated number of beans (x)	Frequency (f)
300–324	9
325–349	26
350–374	52
375–399	64
400–424	83
425–449	57
450–474	18
475–499	5

EXERCISE 40.3H

1 For each of these sets of data, calculate an estimate of
  (i) the range.   **(ii)** the mean.

a)

Height of sunflower in centimetres (x)	Number of plants (f)
$100 \leqslant x < 110$	6
$110 \leqslant x < 120$	13
$120 \leqslant x < 130$	35
$130 \leqslant x < 140$	29
$140 \leqslant x < 150$	16
$150 \leqslant x < 160$	11
Total	110

b)

Weight of egg in grams (x)	Number of eggs (f)
$20 \leqslant x < 25$	9
$25 \leqslant x < 30$	16
$30 \leqslant x < 35$	33
$35 \leqslant x < 40$	48
$40 \leqslant x < 45$	29
$45 \leqslant x < 50$	15
Total	150

c)

Length of green bean in millimetres (x)	Frequency (f)
$60 \leqslant x < 80$	12
$80 \leqslant x < 100$	21
$100 \leqslant x < 120$	46
$120 \leqslant x < 140$	27
$140 \leqslant x < 160$	14
Total	120

d)

Time to complete race in minutes (x)	Frequency (f)
$54 \leqslant x < 56$	1
$56 \leqslant x < 58$	4
$58 \leqslant x < 60$	11
$60 \leqslant x < 62$	6
$62 \leqslant x < 64$	2
$64 \leqslant x < 66$	1
Total	25

2 For each of these sets of data

    (i) write down the modal class.   **(ii)** calculate an estimate of the mean.

a)

Height of shrub in metres (x)	Number of shrubs (f)
$0.3 \leqslant x < 0.6$	57
$0.6 \leqslant x < 0.9$	41
$0.9 \leqslant x < 1.2$	36
$1.2 \leqslant x < 1.5$	24
$1.5 \leqslant x < 1.8$	15

b)

Weight of plum in grams (x)	Number of plums (f)
$20 \leqslant x < 30$	6
$30 \leqslant x < 40$	19
$40 \leqslant x < 50$	58
$50 \leqslant x < 60$	15
$60 \leqslant x < 70$	4

c)

Length of journey in minutes (x)	Frequency (f)
$20 \leqslant x < 22$	6
$22 \leqslant x < 24$	20
$24 \leqslant x < 26$	38
$26 \leqslant x < 28$	47
$28 \leqslant x < 30$	16
$30 \leqslant x < 32$	3

d)

Speed of car in miles per hour (x)	Frequency (f)
$25 \leqslant x < 30$	4
$30 \leqslant x < 35$	29
$35 \leqslant x < 40$	33
$40 \leqslant x < 45$	6
$45 \leqslant x < 50$	2
$50 \leqslant x < 55$	1

3 The table shows the monthly wages of the workers in an office.

Wages in £ (x)	$500 \leqslant x < 1000$	$1000 \leqslant x < 1500$	$1500 \leqslant x < 2000$	$2000 \leqslant x < 2500$
Frequency (f)	3	14	18	5

    a) What is the modal class?
    b) In which class is the median wage?
    c) Calculate an estimate of the mean wage.

4 The table shows the length, in seconds, of 100 calls made from a mobile phone.

Length of call in seconds (x)	$0 \leqslant x < 30$	$30 \leqslant x < 60$	$60 \leqslant x < 90$	$90 \leqslant x < 120$	$120 \leqslant x < 150$
Frequency (f)	51	25	13	7	4

Calculate an estimate of the mean length of a call.

5 The table shows the prices paid for greetings cards sold in one day by a card shop.

Calculate an estimate of the mean price, in pence, paid for a greetings card that day.

Price of greetings card in pence (x)	Frequency (f)
$75 \leqslant x < 100$	23
$100 \leqslant x < 125$	31
$125 \leqslant x < 150$	72
$150 \leqslant x < 175$	59
$175 \leqslant x < 200$	34
$200 \leqslant x < 225$	11
$225 \leqslant x < 250$	5

EXERCISE 41.1H

1   The probability that Stacey will go to bed late tonight is 0.2.
    What is the probability that Stacey will not go to bed late tonight?

2   The probability that I will throw a six with a dice is $\frac{1}{6}$.
    What is the probability that I will not throw a six?

3   The probability that it will snow on Christmas Day is 0.15.
    What is the probability that it will not snow on Christmas Day?

4   The probability that someone chosen at random is left-handed is $\frac{3}{10}$.
    What is the probability that they will be right-handed?

5   The probability that United will lose their next game is 0.08.
    What is the probability that United will not lose their next game?

6   The probability that Ian will eat crisps tomorrow is $\frac{17}{31}$.
    What is the probability that he will not eat crisps tomorrow?

EXERCISE 41.2H

1   A shop has brown, white and wholemeal bread for sale.
    The probability that someone will choose brown bread is 0.4 and the probability that they will choose white bread is 0.5.
    What is the probability of someone choosing wholemeal bread?

2   A football coach is choosing a striker for the next game.
    He has three players to choose from; Wayne, Michael and Alan.
    The probability that he will choose Wayne is $\frac{5}{19}$ and the probability that he will choose Michael is $\frac{7}{19}$.
    What is the probability that he will choose Alan?

3   A bag contains red, white and blue counters. Jill chooses a counter at random.
    The probability that she chooses a red counter is 0.4 and the probability that she chooses a blue counter is 0.15.
    What is the probability that she chooses a white counter?

4   Elaine goes to town by car, bus, taxi or bike.
    The probability that she uses her car is $\frac{12}{31}$, the probability that she catches the bus is $\frac{2}{31}$ and the probability that she takes a taxi is $\frac{13}{31}$.
    What is the probability that she rides her bike into town?

5   A biased five-sided spinner is numbered 1 to 5. The table shows the probability of obtaining some of the scores when it is spun.

Score	1	2	3	4	5
Probability	0.37	0.1	0.14		0.22

    What is the probability of getting 4?

6   A cash bag contains only £20, £10 and £5 notes. One note is chosen from the bag at random.
    There is a probability of $\frac{3}{4}$ that it is a £5 note and a probability of $\frac{3}{20}$ that it is a £10 note.
    What is the probability that it is a £20 note?

EXERCISE 41.3H

1   The probability that United will lose their next game is 0.2.
    How many games would you expect them to lose in a season of 40 games?

2   The probability that it will be rainy on any day in June is $\frac{2}{15}$.
    On how many of June's 30 days would you expect it to be rainy?

3   The probability that an eighteen-year-old driver will have an accident is 0.15.
    There are 80 eighteen-year-old drivers in a school.
    How many of them might be expected to have an accident?

4   When Phil is playing chess, the probability that he wins is $\frac{17}{20}$.
    In a competition, Phil plays 10 games. How many of them might you expect him to win?

5   An ordinary six-sided dice is thrown 90 times. How many times might you expect to get
    a)   a 4?   b)   an odd number?

6   A box contains twelve yellow balls, three blue balls and five green balls.
    A ball is chosen at random and its colour noted. The ball is then replaced. This is done 400 times. How many of each colour might you expect to get?

EXERCISE 41.4H

1   Pete rolls a dice 200 times and records the number of times each score appears.

Score	1	2	3	4	5	6
Frequency	29	34	35	32	34	36

    a)   Work out the relative frequency of each of the scores.
         Give your answers to 2 decimal places.
    b)   Do you think that Pete's dice is fair? Give a reason for your answer.

2   Rory kept a record of his favourite football team's results.
        Win: 32   Draw: 11   Lose: 7
    a)   Calculate the relative frequency of each of the three outcomes.
    b)   Are your answers to part a) good estimates of the probability of the outcome of their next match?
         Give a reason for your answer.

3   In a survey, 600 people were asked which flavour of crisps they preferred.
    The results are shown in the table.

Flavour	Frequency
Plain	166
Salt & Vinegar	130
Cheese & Onion	228
Other	76

    a)   Work out the relative frequency for each flavour.
         Give your answers to 2 decimal places.
    b)   Explain why it is reasonable to use these figures to estimate the probability of the flavour of crisps that the next person to be asked will prefer.

4   The owner of a petrol station notices that, in one day, 287 out of 340 people filling their car with petrol spent over £20.
    Use these figures to estimate the probability that the next customer will spend
    a)   over £20.
    b)   £20 or less.

5   Jasmine made a spinner numbered 1, 2, 3, 4
    and 5.
    She tested the spinner to see if it was fair.
    The results are shown below.

Score	1	2	3	4	5
Frequency	46	108	203	197	96

a) Work out the relative frequency of each
    of the scores.
    Give your answers to 2 decimal places.
b) Do you think that the spinner is fair?
    Give a reason for your answer.

6   A box contains yellow, green, white and blue
    counters.
    A counter is chosen from the box and its
    colour noted. The counter is then replaced in
    the box.
    The table below gives information about the
    colour of counter picked.

Colour	Relative frequency
Yellow	0.4
Green	0.3
White	0.225
Blue	0.075

a) There are 80 counters altogether in the
    box.
    How many do you think there are of each
    colour?
b) What other information is needed before
    you can be sure that your answers to part
    a) are accurate?

42 → PROBABILITY 2

EXERCISE 42.1H

1   A bag contains five red counters, three green counters and two yellow counters.
    What is the probability of selecting
    a)   a red counter or a green counter?
    b)   a green counter or a yellow counter?

2   When Mr Smith goes on holiday, the probability that he goes to the seaside is 0.4, to the countryside is 0.35 and to a city is 0.25.
    What is the probability that he goes on holiday
    a)   to the countryside or the city?
    b)   to the seaside or the city?

3   A spinner is numbered 1 to 5.
    The probabilities of each of the numbers occurring are given in the table.

Number	1	2	3	4	5
Probability	0.39	0.14	0.22	0.11	0.14

    What is the probability that in one spin the number will be
    a)   1 or 2?   **b)**   3, 4 or 5?
    c)   an odd number?   **d)**   less than 2?
    e)   at least 3?

4   There are four kings and four aces in a pack of 52 playing cards.
    A card is chosen at random.
    What is the probability that it is a king or an ace?

5   A coin is tossed and a dice is thrown.
    What is the probability of getting a head on the coin and an even number on the dice?

6   An ordinary dice is thrown three times.
    What is the probability that the dice lands on 6 each time?

7   The probability that the school team wins their next hockey match is 0.8.
    What is the probability that, in their next two matches, the school team
    a)   wins both matches?
    b)   wins neither match?

8   Each of the letters of the word MISSISSIPPI is written on a card.
    The cards are shuffled and one is selected.
    This card is returned to the pack which is again shuffled.
    A second card is selected.
    What is the probability that the two cards are
    a)   both P?
    b)   both S?
    c)   both a consonant?

9   A box contains a large number of red beads and a large number of white beads.
    40% of the beads are red.
    A bead is chosen from the box, its colour is noted and it is replaced.
    A second bead is then chosen.
    What is the probability that
    a)   both beads are red?
    b)   both beads are white?
    c)   one bead of each colour is chosen?

10 In a large batch of light bulbs, the probability that a bulb is defective is 0.01.
    Three light bulbs are selected at random for testing.
    What is the probability that
    a)   all three work?
    b)   all three are defective?
    c)   two of the three are defective?

EXERCISE 42.2H

1 Pat is playing a board game with a spinner numbered 1 to 4.
She spins the spinner twice.
   a) Copy and complete the tree diagram.

First spin	Second spin	Outcome	Probability
Four	Four	FF	
	Not a four	FN	
Not a four	Four	NF	
	Not a four	NN	

   b) Use the tree diagram to work out the probability that
      (i) Pat gets two 4s.   (ii) Pat gets just one 4.

2 The probability that I get up late on any day is 0.3.
   a) Draw a tree diagram to show my getting up late or not late on two days.
   b) Work out the probability that
      (i) I don't get up late on either of the two days.
      (ii) I get up late on one of the two days.

3 There are five red discs and three blue discs in a bag.
A disc is selected, its colour is noted and it is then replaced in the bag.
A second disc is then selected.
   a) Draw a probability tree diagram to show the outcomes of the two selections.
   b) Use the tree diagram to find the probability that
      (i) both discs are red.
      (ii) both discs are the same colour.
      (iii) at least one disc is blue.

4 The probabilities that Phil wins, draws or loses any game of chess are 0.6, 0.3 and 0.1 respectively.
   a) Draw a tree diagram to show the outcomes of Phil's next two games.
   b) Work out the probability that
      (i) Phil wins both games.
      (ii) Phil wins one of the two games.
      (iii) the results of the two games are the same.

5 On the way to work, Beverley passes through three sets of traffic lights.
The probability that the first set of lights is green when she reaches them is 0.6.
The probability that the second set is green is 0.7.
The probability that the third set is green is 0.8.
   a) Draw a probability tree diagram to show the possible outcomes.
   b) What is the probability that she has to stop at
      (i) all three sets of lights?
      (ii) just one set of lights?
      (iii) at least two sets of lights?

EXERCISE 42.3H

1 There are four black beads and three white beads in a box.
A bead is selected at random and not replaced.
A second bead is then selected.
   a) Draw a probability tree diagram to show all the possible outcomes.
   b) Find the probability that the two beads are
      (i) both white.
      (ii) one of each colour.

2 Ben randomly selects three cards, without replacing any, from a normal pack of 52 playing cards.
What is the probability that
   a) all three are aces?
   b) two of the three are aces?

3 The probability of it snowing one day in winter is reported to be 0.2.
If it snows on that day, the probability that it snows the following day is 0.7.
If it doesn't snow, the probability that it will snow the following day is 0.1.
   a) Draw a probability tree diagram to show the possible outcomes.
   b) What is the probability of
      (i) it snowing on both days?
      (ii) it not snowing on both days?
      (iii) it snowing on at least one of the two days?

4 When Elaine goes to school, she either walks, cycles or goes by bus.
The probability that she walks is 0.5 and that she cycles is 0.2.
If she walks the probability that she is late is 0.4, if she cycles it is 0.1, and if she goes by bus it is 0.2.
What is the probability that she is late for school?

5 There are nine boys and fifteen girls in a class. Three children are to be randomly selected to represent the class in a competition.
What is the probability that
   a) all three are girls?
   b) one is a girl and two are boys?
   c) at least two girls are chosen?

6 Rice is sold in large bags. The bag of rice is a mixture of 60% white long grain and 40% black wild rice. Two grains of rice are selected at random.
Calculate the probability that:
   a) Both the grains are black wild rice.
   b) There is at least one grain of black wild rice.

7 A selection box contains 20 toffees. There are 3 with hard centres, 7 with soft centres and 10 with chewy centres. Two toffees are selected at random from the box.
   a) Calculate the probability that both selected toffees have chewy centres.
   b) Calculate the probability that at least one of the toffees has a hard centre.

8 A bag contains 25 wine gums. There are 3 green, 5 yellow, 8 black and 9 red wine gums in the bag. Two wine gums are selected at random from the bag.
   a) Calculate the probability that both selected wine gums are red.
   b) Calculate the probability that at least one of the selected wine gums is yellow.

9 One hundred raffle tickets are sold. The tickets sold are numbered from 1 to 100. The raffle tickets are placed in a drum for a draw. Two raffle tickets are selected, one ticket at a time and not replaced in the drum.
   a) Find the probability that one of the tickets drawn is odd rather than even.
   b) Find the probability that at least one of the tickets drawn is odd.

43 → PLANNING AND COLLECTING

EXERCISE 43.1H

1 State whether the following are primary data or secondary data.
- **a)** Weighing packets of sweets
- **b)** Using bus timetables
- **c)** Looking up holiday prices on the internet
- **d)** A GP entering data for a new patient on his records after seeing the patient

2 Lisa is doing a survey and has written this question.

> What colour is your hair?
>
> Black ☐   Brown ☐   Blonde ☐

- **a)** Give a reason why this question is unsuitable.
- **b)** Write a better version.

3 Steve is doing a survey about his local sports facilities.
Here is one of his questions.

> How much do you enjoy doing sport?
>
> 1   2   3   4   5

- **a)** Give a reason why this question is unsuitable.
- **b)** Write a better version.

4 Mia is doing a survey about school lunches. She gives out questionnaires to the first 30 people in the queue for lunch.
- **a)** Why is this likely to give a biased sample?
- **b)** Describe a better method of obtaining a sample for her survey.

5 Here is one of Mia's questions.

> Don't you agree that we don't have enough salads on the menu?

- **a)** Give a reason why this question is unsuitable.
- **b)** Write a better version.

6 A survey is to be done about school students' earnings and pocket money.
Write five suitable questions which should be included in such a survey.

EXERCISE 44.1H

1 This cumulative frequency graph shows the heights of 80 sunflowers. Find the median, the quartiles and the interquartile range of these heights.

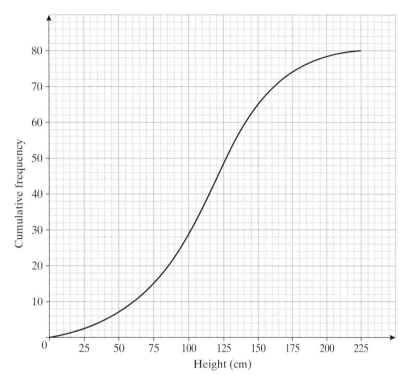

2 The left hand table shows information about the masses of 140 tomatoes.

    a) Copy and complete the cumulative frequency table on the right.

Mass (*m* grams)	Frequency
$0 < m \leqslant 20$	4
$20 < m \leqslant 40$	14
$40 < m \leqslant 60$	25
$60 < m \leqslant 80$	47
$80 < m \leqslant 100$	36
$100 < m \leqslant 120$	14

Mass (*m* grams)	Cumulative frequency
$m \leqslant 0$	0
$m \leqslant 20$	4
$m \leqslant 40$	18
$m \leqslant 60$	
$m \leqslant 80$	
$m \leqslant 100$	
$m \leqslant 120$	

    b) Draw the cumulative frequency graph.

    c) Use your graph to find the median and interquartile range of these masses.

3 The left hand table shows the ages of people in a health club.

Age (years)	Frequency
under 11	32
11–18	25
19–29	53
30–49	83
50–69	45
70–95	21

Age (years)	Cumulative frequency
$y < 0$	0
$y < 11$	32
$y < 19$	57
$y < 30$	110
$y <$	
$y <$	

a) Copy and complete the cumulative frequency table on the right.
   Note: the upper boundary of the 11–18 age group is the 19th birthday.
b) Draw the cumulative frequency graph.
c) How many people in this club are aged under 40?
d) How many people in this club are aged 60 or over?
e) Find the median and the quartiles.

4 The cumulative frequency graph shows the foot lengths of a sample of 50 boys and 50 girls.

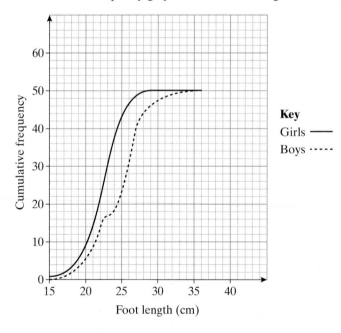

Key
Girls ———
Boys -----

a) What does the flat section at the top of the girls' graph tell you?
b) Compare the distributions. Make two comparisons.

5 This table shows the distance walked by each of a group of students one day.

Distance (d miles)	Frequency
$0 < d \leqslant 2$	4
$2 < d \leqslant 4$	16
$4 < d \leqslant 6$	8
$6 < d \leqslant 8$	6
$8 < d \leqslant 10$	2

Draw a cumulative frequency graph to represent this distribution.

6 The cumulative frequency graph represents the distances swum by children in a sponsored swim.

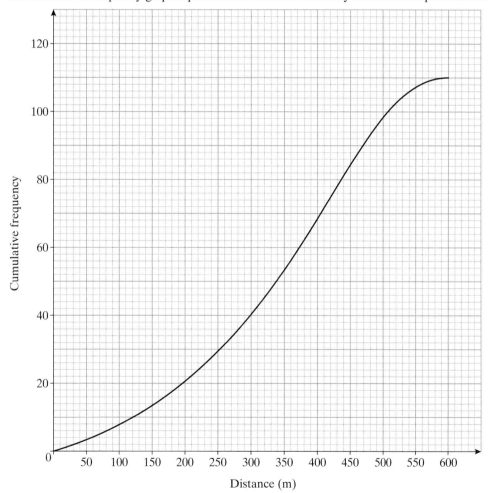

a) How many children took part in the swim?
b) How many children swam over 400 m?

7 A company tested a sample of 200 torch batteries of each of two types it produces.
The table summarises the results, showing the time, in hours, that each battery lasted.

a) On the same axes, draw cumulative frequency graphs to represent these distributions.
b) Which of the two types of battery is more reliable?

Time (t hours)	Frequency for type A	Frequency for type B
$0 < t \leqslant 5$	4	12
$5 < t \leqslant 10$	31	61
$10 < t \leqslant 15$	45	71
$15 < t \leqslant 20$	87	32
$20 < t \leqslant 25$	27	16
$25 < t \leqslant 30$	6	8

EXERCISE 44.2H

1 The table shows the amounts spent at the supermarket by a sample of people.

Amount spent (£s)	$0 < s \leqslant 20$	$20 < s \leqslant 40$	$40 < s \leqslant 70$	$70 < s \leqslant 100$	$100 < s \leqslant 150$
Frequency	12	16	33	12	6

Draw a histogram to represent this distribution. Label your vertical scale or key clearly.

2 This distribution shows the ages of people visiting a swimming pool one day.

Age (years)	Under 10	10–19	20–29	30–49	50–89
Frequency	96	58	36	58	144

a) Explain why the boundary of the 30–49 group is 50 years.
b) Calculate the frequency densities and draw a histogram to represent this distribution.

3 The histogram represents a distribution of waiting times for non-emergency operations at a particular hospital.

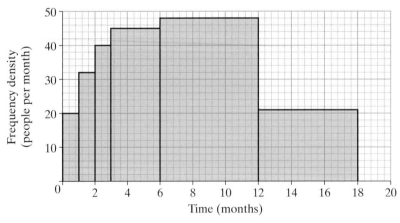

a) Make a frequency table for this distribution.
b) Calculate an estimate of the mean waiting time.

4 The histograms represent the times spent on a piece of coursework by a sample of girls and boys.

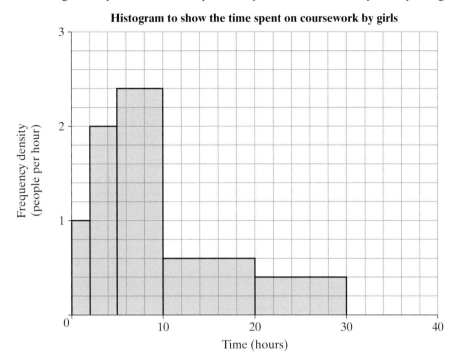

Histogram to show the time spent on coursework by girls

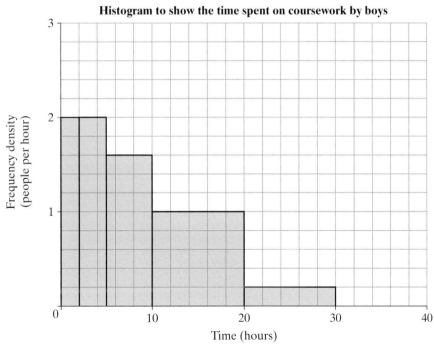

Histogram to show the time spent on coursework by boys

a) Find how many girls and how many boys spent between 5 and 10 hours on the coursework.
b) Compare the distributions.

5 The members of a gym were measured. The histogram represents their heights.

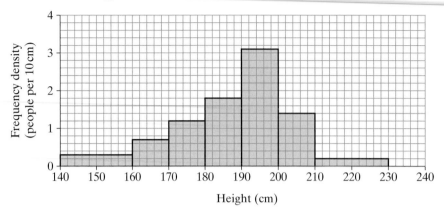

a) How many members of the gym were measured?
b) Calculate an estimate of their mean height.

44 Representing and interpreting data